Smithsonian
Ocean

To stand at the edge of the sea, to sense the ebb and flow of the

tides, to feel the breath of a mist moving over a great salt marsh,

to watch the flight of shore birds that have swept up and down

the surf lines of the continents for untold thousands of years,

to see the running of the old eels and young shad to the sea, is to

have knowledge of things that are as nearly eternal

as any earthly life can be.

— RACHEL CARSON

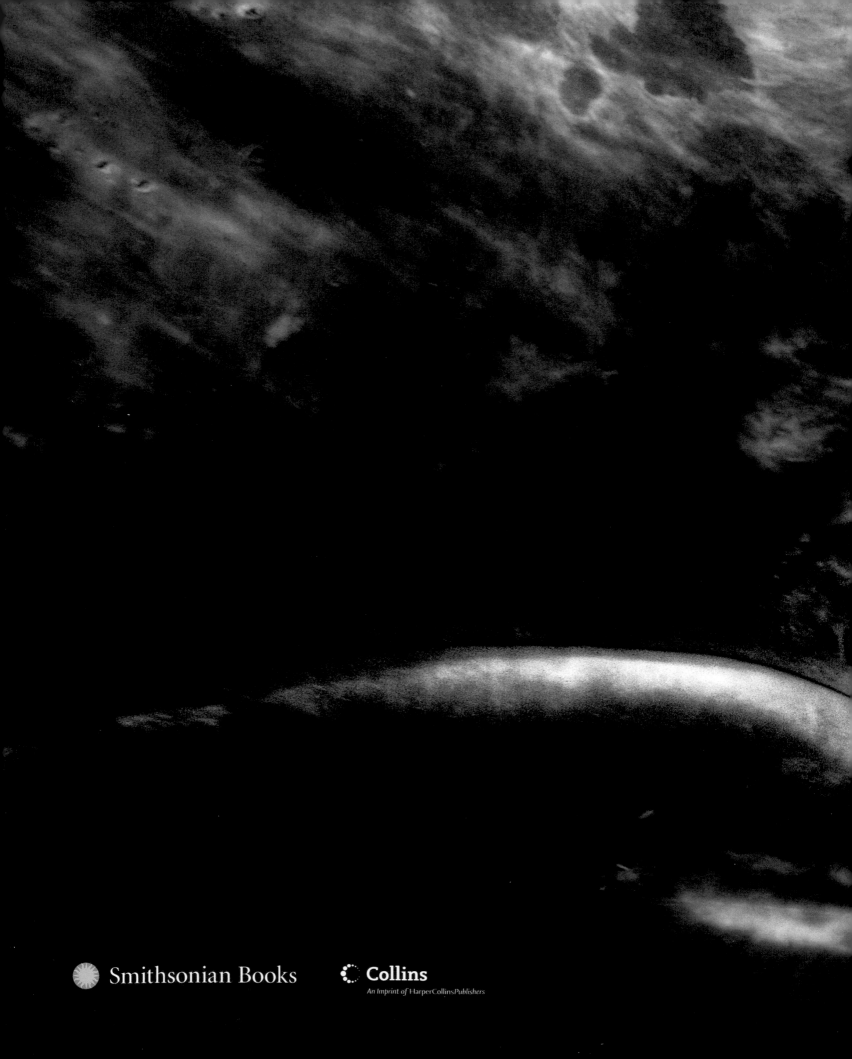

Smithsonian Books

Collins
An Imprint of HarperCollinsPublishers

Smithsonian Ocean

Our Water
Our World

DEBORAH CRAMER

Front cover: Male humpback whale sings off the coast of Maui, Hawaii.

Front endsheet: Leatherback hatchling makes its way to the sea in Tamarindo Bay, Costa Rica.

Pages 2–3: Salt marsh, the Netherlands.

Pages 4–5: Black skimmers, Nickerson Beach, New York.

Pages 6–7: Glass eels, Kennebec estuary, Maine.

Title page: Gray whale, San Ignacio Lagoon, Mexico.

Back endsheet: Mother polar bear and her yearling, Wapusk National Park, Manitoba, Canada.

Back cover: Elegant terns near Isla Rasa in the Sea of Cortez.

SMITHSONIAN OCEAN
Our Water, Our World

Published on the occasion of the opening of the Ocean Hall at the Smithsonian's National Museum of Natural History, September 2008. The National Oceanic and Atmospheric Administration (NOAA) is a contributor to the Ocean Hall.

First edition published in 2008 in the United States of America by Smithsonian Books in association with HarperCollins Publishers.

HarperCollins books may be purchased for educational, business, or sales promotional use.
For information please write:
Special Markets Department
HarperCollins Publishers, 10 East 53rd Street, New York, NY 10022.

Except where otherwise noted, all mapping in this book is generated from Collins Bartholomew digital databases. Collins Bartholomew, the UK's leading independent geographical information supplier, can provide a digital, custom, and premium mapping service to a variety of markets.

For further information: Tel: +44 (0) 141 306 3752
E-mail: collinsbartholomew@harpercollins.co.uk
Or visit: www.collinsbartholomew.com

Pages 2–7 quotation from Rachel L. Carson, *Under the Sea-Wind: A Naturalist's Picture of Ocean Life* (New York: Simon and Schuster, 1941), p. xiii.

Executive Editor: Caroline Newman
Designer: Service Station | Bill Anton
Editor: Duke Johns
Editorial Assistant: Megan Miller
Production Manager: Karen Lumley
Photo Research: Nature's Best Publishing LLC
    Photo Editors, Miriam Stein and Bob Tope
Cartography: Collins Bartholomew, Glasgow, UK

Printed in China

Library of Congress Cataloging-in-Publication Data
Cramer, Deborah.
Smithsonian ocean : our water, our world /
    Deborah Cramer.
    p. cm.
Includes bibliographical references.
ISBN 978-0-06-134383-4 (alk. paper)
1. Marine biology. 2. Marine ecology. I. Title.
QH91.C73 2008
578.77–dc22   2008015633

08 09 10 11 12 SCP 10 9 8 7 6 5 4 3 2 1

Tarpon and jack, Florida Keys National Marine Sanctuary.

Contents

Touched by the Sea

Contents pages: Kelp forests off the coast of California.

Touching the Sea

Foreword

CRISTIÁN K. SAMPER

Director, National Museum of Natural History

An Adélie penguin poised on the ice at Brown Bluff on the Antarctic Peninsula.

Smithsonian Ocean, a remarkable journey through space and time, explores the vital relationship between life and the sea. Deborah Cramer, a gifted and passionate writer, has created a scientifically grounded and elegantly written book, a major contribution to an increasingly important field.

The ocean covers more than two-thirds of the surface of our planet, yet both scientific and public understanding of the ocean are limited. That is why the Smithsonian's National Museum of Natural History has created a new permanent Ocean Hall, the most ambitious renovation in the museum's history. *Smithsonian Ocean* stands alone yet also serves as a powerful exploration of the exhibition's overarching theme: the ocean is a global system essential to all life, including yours. The volume expands on other exhibition concepts as well, including the rich diversity of ocean habitats; the interconnections between the ocean, other global systems, and our daily lives; and the recent impact of humans on the ocean that created life and sustains it still.

Much of what we know about the ocean comes from museum scientists such as those at the National Museum of Natural History and other Smithsonian research centers. Many scholars are generating new knowledge about the ocean by working with Smithsonian marine collections, the world's largest, with over 50 million specimens including fish and invertebrates, fossil plants, animals, and unicellular organisms, many on display in the new Ocean Hall and photographed especially for this book.

I encourage you to read Cramer's volume and enjoy her inspiring narrative and its spectacular images. All those who dive deeply here will emerge enriched by the experience.

Introduction

Many marine animals depend upon each other. Here, juvenile cardinalfish find shelter within the spines of a sea urchin.

Ever since the first rain fell to fill earth's ocean, the sea has circled our planet in an unending flow of life-giving water. While comets sprinkle the planet with "fresh" water, and some new water seeps into the sea from hot springs on the ocean floor, most of earth's water is four billion years old, as old as the earth itself.

All the water we will ever have, it is our water, and it has made our world. Our lives, and the lives of almost all that dwell on earth, depend on the sea.

Tiny cells of chlorophyll float in sunlit waves at the sea surface. Fossils of odd animals, long extinct, that dwelled in a sea long gone, now lie in high mountains. These plants and these animals, living and dead, these mountains, those rising and those crumbling away, and this enduring sea: in their story lies our own.

It is a story that begins in the ocean.

There life emerged and evolved. The first living cell was born in the scalding waters of deep-sea hot springs, whose very existence overturned our long-held assumptions about the ingredients necessary for life. A nurturing sea gave rise to earth's first cell, first plant, and first animal. We are, essentially, specialized fish.

Evolution took place not only as a result of genetic innovation, but within a context of opening and closing ocean basins, drifting continents, rushing currents, advancing and retreating glaciers, massive volcanic eruptions. This history, recorded in remnants of ancient seafloor, reveals how earth's first photosynthesizing organisms lived in the sea; how, little by little, they filled the atmosphere with oxygen, ushering in the dawn of animal life; how closing ocean basins created landscapes that made it possible for fish to come ashore; how changes in climate wrought by the sea helped guide the evolution of *Homo sapiens*.

The sea is both a cradle and a museum of seemingly infinite forms of life still largely unnamed and unknown to science. *Smithsonian Ocean*, reflecting and expanding upon the major themes of the National Museum of Natural History's new Ocean Hall, explores the sea's dazzling diversity and examines the myriad ways the sea touches us, and we, increasingly, touch the sea.

Living in a human-made world, we have lost sight of how much we depend on the ocean. *Smithsonian Ocean*, a journey through that ocean and through time, plumbs the depths of the vital partnership between life and the sea that sustains us. Parts One and Two, "Beginnings" and "Touched by the Sea," explore the sea's essential role in that partnership. Pairing each chapter in these sections with a bridge to the ancient ocean suggests how profoundly earth's past creates and illuminates the present, and anticipates a future.

The sea still holds us, spawning and withholding rains that bring about the rise and fall of cities and civilizations. Water evaporated from the sea helps endow our planet with a benign climate, and rushing currents churn up nutrients that create bountiful fisheries. Yet its health is now increasingly in our hands.

Atlantic spotted dolphins socializing in the waters of the Bahamas.
Seen and unseen in every food web is the vital partnership between life and the sea.

Part Three of *Smithsonian Ocean*, "Touching the Sea," considers how the arrival of humans late in earth's biography has shifted the balance between life and the forces that created it, and what that might mean in light of our deep need for and connection to the sea. Our contribution to this partnership is being recorded on the seafloor and will remain, long after we are gone. We choose our legacy together.

All of us, wherever we live, are joined by flowing water, and all the waters of the sea, whether they be at the icy poles or in the clear warmth of the tropics, are one.

The partnership between life and the sea began more than three and a half billion years ago. A window into those early days lies not near shore, that sea edge with which we are most familiar, but in deeper, more distant water where new seafloor is born.

Heat rising from deep within the earth creates new seafloor, new land, and the cradle where life began.

Beginnings

Eden in the Depths

Nine Degrees North, East Pacific Rise, is an unassuming address in the Pacific Ocean. The name barely describes the place, indistinguishable on the broad expanse of unbroken sea. To the person gazing from a ship, the water—warm, clear, and seemingly empty—gives no hint of the strange and exotic world concealed in the depths. The near freezing, sunless realm of the deep sea has long been considered one of earth's most hostile environments, inimical to life itself. Yet, 400 miles (640 km) off the coast of Mexico, on the seafloor at Nine North, there lies a lush and vibrant oasis teeming with animals. To the uninitiated, the environment is forbidding and the animals odd, but it was here in the hot springs of the deep sea that earth's first life may have emerged.

Nine North can be a violent place, wracked by earthquakes and erupting volcanoes. Here the planet is giving birth; its youngest seafloor is being formed. Draped over the young rocks are mats of bacteria whose ancestors are almost as old as the earth itself. Young and old are partners here at Nine North, and at hundreds of other hot springs found throughout earth's ocean. There earth's primordial life forms and its youngest, newest ocean floor together nourish communities that continually shatter our assumptions about what sustains life. They are a living reminder of how little we really know the waters covering two-thirds of our planet, and how much life and the forces that give birth to the ocean are linked.

Metal sulfides precipitate out of scalding vent water, building tall vent chimneys at Nine North in the East Pacific Rise.

Pulses in the Deep

Geology is the foundation: in the hot springs of the deep sea, life emerges, literally, from rock. The process began early in earth's history, when the surface of the cooling planet cracked. Today earth's crust is broken into approximately seven giant pieces and a host of smaller ones. The plates drift, carried by heat from deep within earth's core. Where they move apart, molten rock rises, building chains of black, undersea volcanoes that circle the globe like the seams of a baseball. They are the longest mountain range on earth, but few people have scaled their summits or walked their valleys. Mostly unseen and unheard, mid-ocean ridges account for 90 percent of earth's volcanic activity. From them the sea is born.

Nine North sits astride these mountains of the sea in the East Pacific Rise, at the border between the Pacific Plate, earth's largest, and the Cocos Plate, one of the smallest. As these plates separate, by 4 to 4½ inches (11 to 12 cm) each year, molten lava rises to fill the gap, building the new edge. The eruptions and accompanying earthquakes are invisible beneath the waves, muffled by the weight of water, but seismometers resting on the seabed take their pulse. It is unsteady.

Opposite: A garden of giant tubeworms brightens deep-sea hot springs in the East Pacific Rise.

Below: A vent crab, Bythograea thermydron, *lives among the tubeworms and on mussel beds at deep-sea hydrothermal vents.*

Nine North trembles, struck by two or three earthquakes each day, but sometimes its heartbeat quickens. Beginning in 2003, first tens and then hundreds of earthquakes cracked the seafloor daily, hinting at an impending eruption. Scientists monitoring the site in the spring of 2006 were unable to retrieve eight of twelve seismometers. The data they did recover explained why. Beginning at 13:45 Greenwich mean time on January 22, as many as 250 earthquakes wracked the seafloor each hour. Lava welled up out of the depths, spreading over 1.1 mile (1.8 km) and burying the seismometers, each the size of a microwave oven, in earth's youngest terrain.

Tubeworm Barbecue

More was buried than scientific instruments. By sheer luck, in 1991 scientists had witnessed the immediate aftermath of an earlier volcanic eruption at Nine North, their

first on a mid-ocean ridge. Peering from the portholes of *Alvin*, a deep-diving submersible whose titanium shell protected them from the crushing pressure at 8,270 feet (2,520 m), they saw the newborn seafloor: only days before, within the space of about two hours, approximately 400,000 truckloads of lava had been added to the seabed. Death and devastation accompanied creation. The young seafloor was fresh and glassy, but the animals were dead and dying. Mussels were shattered, blown apart by the eruption. Bits and pieces of dismembered tubeworms, scorched and charred by the heat, littered the seabed. A partially cooked limpet lay near traumatized tubeworms whose days were numbered. The kill was so fresh that the victims had yet to decompose, and scavenging crabs had yet to arrive. Scientists nicknamed the site the "Tubeworm Barbecue."

See Map 3, Hydrothermal Vents, p. 266.

Life Reemerges from the Ruins

Death, despite its gruesome appearance, did not linger. The upheaval that builds an ocean also renews and replenishes life, delivering an infusion of energy-rich water. The source of this elixir is a chemical reaction between rock and water, made possible by the moving earth. The reaction begins as icy seawater percolates through young, cracked seafloor, heats up, and turns acidic. At 660 to 750 degrees Fahrenheit (350–400°C), it leaches copper, iron, and zinc from the surrounding rock, becoming a corrosive liquid barely resembling water. Buoyant, and laden with metal, the scalding jet of fluid rises and breaks through the seabed. As it hits the chilly water, metals rain out in billows of black smoke, constructing chimneys and towers around the vents. From this poisonous brew, life emerges, and scientists at Nine North bore witness.

Black smokers belch scalding 626-degrees-Fahrenheit (330°C) fluid from hydrothermal vents in the mountains of the Mid-Atlantic Ridge, where the seafloor splits apart.

The setting was more Hades than Eden. The temperature in the "black smokers," as they are called, had risen to 757 degrees Fahrenheit (403°C), and the vent water was laced with brimstone. Toxic to humans, this hydrogen sulfide is manna from heaven for bacteria that spewed from chimneys and cracks in the seafloor, shrouding *Alvin* and its passengers in a blizzard of white. Feeding on hydrogen sulfide, the bacteria quickly grew into thick mats carpeting the young rock. Unknown, unnamed, they would become the foundation for life that would rise and flourish from the destruction. Scientists called the site "Phoenix."

The Staff of Life

Scientists returned to Phoenix year after year and watched the oasis bloom. Limpets, crabs, mussels, and giant tubeworms thrive in the shimmering water around the vents. Normally, life in the deep sea is hard, and animals few and far between, but this garden is luxuriant. As many as 600 tubeworms can crowd in a single cluster. White giants tipped with blood-red plumes, each worm lives in a tube about 4.5 feet (1.4 m) long. Fat and healthy, growing as much as 33 inches (85 cm) a year, they are among the fastest-growing invertebrates of the sea. The worms are obviously well nourished, even though they lack mouths and digestive systems.

Bacteria residing within the tubeworms prepare the sumptuous repast. Tubeworms and their bacterial partners engage in a symbiotic relationship that has transformed our understanding of biology, unveiling an entirely new realm of existence. Up on the surface of the sea and on dry land, plants use energy from sunlight to make carbohydrate consumed by animals. At deep-sea hot springs, chemosynthesis, not photosynthesis, is the foundation of life. Bacteria living in the tubeworms use hydrogen sulfide from vent water rather than sunlight to make food for their hosts.

In other circumstances, this endeavor would be lethal. The high hydrogen sulfide concentrations here would poison and kill most animals, but tubeworms thrive on the potentially deadly combination of gases. Only now are scientists learning how. Using X-ray crystallography to study tubeworms at Nine North, they discovered that as a worm's plume takes in hydrogen sulfide, zinc in its hemoglobin temporarily binds it,

preventing the fatal reaction. Essential nourishment is thereby delivered to the bacteria without killing its host.

The plenty reaches beyond an unusual worm gorging in the darkness of a distant ocean. The protective hemoglobin, an exciting discovery in marine biology in and of itself, is now the subject of medical research. It serves as a model for an artificial blood substitute in transfusions, one of only many examples where human need may be fulfilled by the sea's vast, and still mostly undiscovered, rich diversity.

Scientists tracking the emergence of life at Nine North have been generously rewarded. Watching tubeworms mature and spawn, and recovering their larvae, researchers now understand the beginnings of this life-sustaining bacterial partnership. Surprisingly, it starts with a skin infection. Young tubeworms don't ingest the bacteria, as scientists first assumed. Rather, the tiny microbes enter through the worms' skin, migrating through layers of tissue to what will become a large and spacious home, the trophosome. When it is built, and the symbiotic bacteria are firmly established, the trophosome of an adult tubeworm houses billions of bacteria that provide their host with an endless supper.

A Faint Light

Although the deep sea's hydrothermal vents, far removed from the sight of land dwellers, seem a world apart, slender threads tie them to the rhythm of life at the surface. Currents sinking in polar seas carry in oxygen required by symbiotic bacteria. Juvenile shrimp floating through the water before settling at a vent survive on tiny plants and animals sinking from the sea surface. Sunlight is key: photosynthesis produces oxygen and food to feed young shrimp.

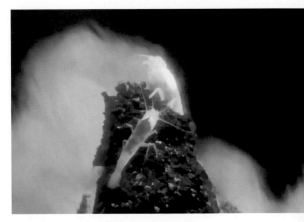

Photosynthesis takes place only in sunlight—or so biologists thought. No one imagined it could take place in the blackness of the deep-sea vents. Like the vent life itself, that seemed impossible until a marine biologist discovered a light-absorbing patch on the backs of "blind" shrimp swarming at an Atlantic vent field. This vestige of an eye detects dim light emanating ever so faintly from the glowing heat of the vents, from the flash of imploding gas bubbles, from the fracturing of mineral crystals.

Opposite and right: Eyeless shrimp swarm in the warm water between the scalding vent fluids and the freezing deep sea.

Scientists, seeing light, sought a photosynthesizing organism. After much searching, they found, in fluid pouring from a vent at Nine North, a tiny, rod-shaped, green sulfur bacterium. Though barely alive and growing ever so slowly (dividing perhaps once every two or three years), the very existence of GSB1, as it is called, further expands our understanding of biology. GSB1 may be the first known organism to make energy from light other than the sun, and it may offer an evolutionary link to the origins of photosynthesis.

A New Sea Is Born

New seafloor is added to ocean basins at the mid-ocean ridges. Up on dry land, East Africa is breaking apart to give birth to a new ocean. The rising heat uplifts, stretches, and cracks the land, sculpting Mount Kilimanjaro, the plains of the Serengeti, and the sparkling lakes of Kenya and Tanzania. Land has already given way to ocean where Africa and the Arabian Peninsula split, letting in the Red Sea. Africa and Arabia separate in fits and starts: in less than a week during September 2005, the desert of Afar, Ethiopia, the youngest segment of this young sea, widened by 26 feet (8 m).

As this hot, bleak place is ripped apart to create a new sea, ancient fossils long buried are uncovered. Afar records two births: the birth of an ocean and the birth of the human race. Thousands of fossils in Afar are yielding the story of our own evolution, including three-million-year-old bones of our walking ancestors—the hominid Lucy and the skeleton of a three-year-old girl—and even older bones marking the divergence of humans and chimpanzees. Tools and animal bones lie in the sand and silt where our forebears hammered tools from volcanic rock formed by the opening sea, and then dismembered antelope and three-toed horses living at the water's edge.

Beyond the Volcanoes of the Ridge

Rising from an undersea mountain in the Atlantic are hundreds of ghostly white limestone spires—the Lost City. To scientists who'd come to map the mountain, the slender, graceful chimneys and giant beehive-shaped mounds were a surprise. No lava flowed here. The rocks, one and a half million years old, had cooled long ago. The water shimmering near the vents was heated by another source.

As seafloor moves away from the ridge, it stretches, thins, and cracks, exposing mantle rock from deep within the earth. The mantle rock reacts with seawater, producing heat. Lost City is self-sustaining. As the mantle rock absorbs water, it swells, fracturing more rock and enabling more water to percolate through to generate more heat. Lost City is 30,000 years old: the partnership between rock and sea could power this field for hundreds of thousands, perhaps millions, of years.

Warm, alkaline water gently diffuses from the chimneys at Lost City, where a pinch of soft, crumbly limestone harbors between 10 and 100 million microbes, testament to the tenacity of life in seemingly hostile environments. Living and breathing methane, they open yet another window into the past, another clue to earth's early life.

Away from the vents and closer to shore, methane and oil, seeping from the seabed through cracks in the rock and sediments, support similar communities of tube-worms, clams, and mussels. Among the approximately two hundred species of animals living at cold seeps is a tubeworm with a 250-year life span. Recycling may explain this extraordinary longevity. The tubeworm, *Lamellibrachia luymesi*, releases waste sulfate through its roots. Bacteria in the sediment then convert it into sulfide that feeds the worm.

In this digital composite, the Atlantic Ocean's Lost City vent field, whose towers reach as high as 200 feet (60 m), dwarfs the research submersible Alvin.

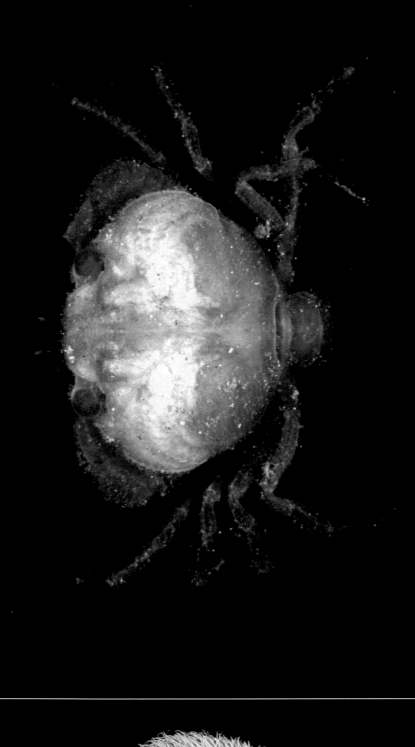

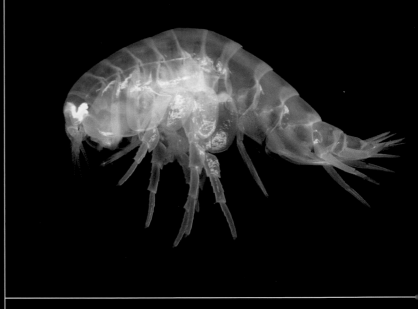

The discovery at the sea's deep hot springs that water, not light, is essential to life opens the possibility that life may exist or may have existed beyond earth: in the gullies and canyons of Mars where water once flowed, in its thick polar ice cap, and in the deep liquid sea beneath the frozen surface of Jupiter's moon Europa.

Continual Revelation

Not that long ago, deep-sea hot springs were thought to be anomalies, odd and rare occurrences. Instead they are the multitudinous offspring of a shifting, restless earth. Oases in the dark sea, they teem with life: gleaming white clams as large as dinner plates, swarming shrimp, thick mussel beds, gardens of giant tubeworms. Scientists exploring the sea's hydrothermal vents have found over 500 previously unknown animal species. As new sites are visited, the number increases.

The Kairei vent field in the Indian Ocean revealed a one-of-a-kind snail. Unlike any other mollusk known today, it resembles armored animals—halkieriids and tommotiids—that first appeared in the sea 500 million years ago. Black, magnetic scales, overlapping like roof tiles and coated with iron sulfide, cover its foot. Perhaps they are built by symbiotic bacteria. Perhaps they serve to stymie predators whose venomous darts cannot pierce the scales. Whatever their source and whatever their purpose, their presence in the modern ocean suggests both that old evolutionary adaptations are not forgotten and that those that do appear reflect their environments. A hydrothermal vent, rich in iron sulfide, is a fitting place for an iron-clad snail.

South of Easter Island, at mid-ocean ridge vents known as Sebastian's Steamer, Pâle Étoile, and Annie's Anthill, scientist found an unusual hairy crab amid the vent mussels, basking in the warm water. They scooped it up with a slurp gun and, because the shaggy tufts of hair on its legs reminded them of the Himalayan abominable snowman, informally called it "yeti." Realizing upon closer examination that their crab belonged not only to a new species but also to a new genus and a new family, they called it *Kiwa hirsuta: Kiwa* for the goddess of shellfish in Polynesia, and *hirsuta* for its long, pale hair.

Opposite: Deep-sea hot springs support a variety of residents. Left above: A vent crab, Bythograea thermydron. *Left below: Deep-sea Pompeii worm,* Alvinella pompejana, *lives on vent chimneys. Top right: Deep-sea amphipod. Bottom right: A "yeti crab,"* Kiwa hirsuta.

Above, top: This deep-sea shrimp is a visitor to the hydrothermal vents. Above, bottom: Scale worms have been found living on the walls of vent chimneys and inside vent mussels.

Scientists saw the crab picking at cracked mussels, but they also suspect that its hair may be an important source of nutrition. Mats of bacteria, soaking up the plentiful hydrogen sulfide, may have colonized the hairs, providing a convenient and large meal. So much about this crab is unknown. Does it live only at the vents south of Easter Island, or does it also dwell further south in hot springs beneath the high winds and fierce storms churning Antarctic waters? Only further voyages of exploration will yield an answer.

Even the chilly waters of the poles are home to hydrothermal vents. High in the Arctic, between the islands of Iceland and Svalbard, lies the Gallionella Garden. Named for the bacteria covering the site, it is the most northern vent known—so far. Only 10 percent of the ridge has been probed for deep-sea hot springs; there's no telling what future expeditions will reveal.

Increasingly sophisticated technology offers more insight into the diversity at deep-sea hot springs. New techniques of genomic analysis applied to samples taken from vents at the Axial Seamount, a volcano along the mid-ocean ridge west of Oregon, found thousands of new microbes that had never been catalogued before. Until we have identified every possible way to "see" into the ocean, we may never realize the breadth of its life.

The lush oases of deep-sea hot springs teem with life in an environment that can only be described as extreme. The shortest of distances separates the near-freezing water of the deep sea from scalding hydrothermal fluid. The newly visited Turtle Pits vent field lies near Ascension Island, just south of the equator in the Atlantic. There, where the sea is widening as the African and South American plates drift apart, boiling vent water vaporized at 407 degrees Centigrade (765°F), the highest water temperature ever recorded in a hydrothermal vent.

Opposite: The hydrothermal vent fish, Thermichthys hollisi *, swims near Nine North. About 8.5 to 12 inches (218–304 mm) long, and with a blunt snout, it has been seen eating other vent fish half its size*

Top right: At least sixteen species of Munidopsis, *a squat lobster, are now known from vent fields throughout the Atlantic, Pacific, and Indian oceans.*

Below right: Cold-seep tubeworms, Lamellibrachia luymesi, *can live as long as 250 years.*

Lifelines

Slowly, the vents are revealing their secrets. Still, there is much to learn. The distance between hot springs is great, seemingly insurmountable for tiny larvae, yet the offspring of vent animals disperse through the ocean and settle at other hot springs. Their routes are poorly understood: they may catch rides on vent water.

Scientists working at the Indian Ocean's Carlsberg Ridge, finding their water samples clouded, discovered hydrothermal fluid surging from a volcano beneath their ship. Spreading over 43 miles (70 km) and rising 4,600 feet (1,400 m) off the seafloor, the cataclysm shot 1,089 cubic miles (4,540 km^3) of water into the sea, as much as a more typical vent site releases in a year. Plumes emanating from the sea's hot springs may lift larvae up into oceanic currents. There they may drift to new homes, enabling vent animals to survive generation after generation, even though the vents themselves are ephemeral.

Below: Stalked barnacles (Neolepas sp.) live at vents throughout the ocean.

Opposite: Limpets, brittle stars, and a worm crowd on a mussel at the East Pacific Rise.

Decomposing whale skeletons and cold seeps slowly leaking oil and gas onto the seafloor also support communities of life fueled by hydrogen sulfide and methane. They, along with other vent sites awaiting exploration, may be important stepping stones for larvae, enabling animals to colonize new vents. Mussels, for example, first lived in shallow water and then diversified, moving deeper and deeper, arriving first at decomposing whale bone and then at hot springs. Clams, too, originated in shallow water, while tubeworms appear to be vent natives.

No matter their origin, or the route they take between vents, the animals that inhabit today's deep-sea hot springs are young. They evolved recently, within the last 100 million years, while the foundation of life in the vents, the bacteria, are much, much older. Their lineage extends back to the origin of life on earth.

First Land, First Sea

If earth's history were compressed into one year, *Homo sapiens* appeared in the last minute of the last hour of the last day, a late arrival in an expanse of time that encompasses hundreds of thousands of millions of years. Throughout this virtually unfathomable passage, ocean basins opened and closed, continents fused and broke apart, and whole groups of animals arrived and departed. Scaling earth's history into a year, reptiles appeared in the first part of December, and dinosaurs became extinct on Christmas. Throughout it all, microbes endured.

Highly versatile, microbes have persisted for at least three and a half billion years of earth's history, surviving the cataclysms and catastrophes that repeatedly annihilated other forms of life. They were the first life to appear on the young earth; in all likelihood, they will remain when we are gone. While they are maligned, often misunderstood merely as germs, they are the foundation of life, setting the stage that made our entrance possible.

Born from lifeless gasses and molecules in earth's early ocean, they grew in numbers and eventually reconstructed the entire atmosphere, imbuing it with the oxygen we require to live. They evolved, greatly growing in size and complexity, eventually becoming the animals that would become our forebears. While appearances may suggest we have little in common with these invisible microbes, we share a common ancestor. From their story comes ours.

A Hellish Start

The record of earth's beginnings has vanished, its early sea and scraps of continent cycled and recycled to make new ocean basins and new continents. The still, lifeless surface of the moon (little changed since its chaotic origin), the ancient craters of Mars, and the bubbling lava domes on Venus hint at that early time approximately 4.6 billion years ago, when clumps of dust and debris swirling around the sun congealed into rocky chunks that in turn would collide and take the shape of planets.

As the earth grew, it contracted and its interior melted. Radioactive elements in the core began to decay, releasing heat. Meteors cauterized its surface, turning it into a sea of liquid rock and erasing its history. Eventually the bombardment slowed, the planet's fever subsided, and molten lava began to cool into rock. Carbon dioxide

Background: The early earth may have resembled this lava mountain on Mars.

Opposite, above: Heat and lava poured forth from the earth's interior in its youth. The upheaval continues today (here on Isabela Island, Galápagos).

and water vapor emitted from volcanoes and carried in on comets began forming an atmosphere. Water fell as rain but turned to steam on scorching rock. The sea was hot, nearly boiling. Earth's excruciatingly hot and violent beginning lasted for 600 million years, leading scientists to call it the Hadean eon, evoking Hades, the Greek underworld.

THE LAND BESIDE THE SEA

The remains of lands that once sat beside earth's early sea have been crushed and deformed, exposed and eroded, as earth's internal heat engine pushes continents together and breaks them apart, reshaping the planet's contours. They lie scattered throughout the world: in the rugged mountains of Barberton, South Africa, at the foot of ice sheets in southwestern Greenland, and on the shores of the Acasta River in Canada's Northwest Territories. Over the years they have been compressed and baked by moving continents, scoured by glaciers, and weathered almost beyond recognition, but their original identity persists: crystals locked in the ancient rocks of Canada are four billion years old.

Earth's first lands were built from the rock of seafloor, from volcanic lava that rose above the water. Mostly they are gone, but today Iceland, an island also born of the sea, wracked by volcanoes, may echo that early time. It is a fiery land, a piece of the mid-ocean ridge rising above the waves. In the sixth century, Irish monks arriving at its shores in wicker boats saw boiling seas and mountains spewing flaming rock, and assumed they had reached the gates of hell. Iceland's fires continue to burn. Eruptions have swallowed farms and fields, burned crops, and buried homes and businesses as the volcanoes beneath the island continue to sculpt new seafloor.

Early in earth's history, tiny continents and seafloor began to drift, colliding and dividing. Northeast of Reykjavik in the valley of Thingvellir is a deep cut where the land is being wrenched apart today. Thingvellir, the site where Iceland's parliament first met in A.D. 930, lies in a valley edged by steep cliffs. On one side, the land is being pulled toward Europe; on the other, to America.

Thingvellir, Iceland, is being pulled apart: one side drifts toward Europe, the other toward North America.

Another Hellish Start

Eventually earth's surface cooled a little more, leaving a few islands of dry land in a sea filled with rain. Early earth contained the seeds of life: water in an ocean; minerals in hot, molten rock rising from the planet's core; carbon dioxide and nitrogen released from volcanoes. How and under what conditions life emerged from these elements are still mysteries. Scientists have yet to duplicate the process in a lab, and have yet to pinpoint the time and place of that momentous transformation.

Perhaps the first vital force was replication—a strand of RNA reproducing itself again and again. Perhaps it was metabolism—an organic compound producing or capturing energy to sustain itself. Or perhaps it was isolation—organic compounds separating from their surroundings within a membrane, the forerunner of a cell wall.

Whatever that first force, it most likely took place in the ocean. It could have occurred, as Charles Darwin suggested, in a "warm little pond" where a flash of lightning lit the spark of life. Perhaps life began closer to shore, where a wave pushed a fatty hydrocarbon into a tiny bubble that would become a cell. Or perhaps an ancient deep-sea hot spring, laden with minerals, held the cradle of life. Heat rising from within the earth, the same heat that builds continents and opens ocean basins, could have both shaped our planet and endowed it with life. Predecessors of the deep-sea hot springs that give life to Nine North, Lost City, and other hydrothermal oases on the seafloor today could have spawned earth's first living organism.

Opposite and above: Heat-loving bacteria, like those in Yellowstone's Grand Prismatic hot spring, lived in the ocean's ancient hydrothermal vents at the dawn of life.

It may have begun with a simple chemical reaction. Hydrogen and iron sulfide carried in hot hydrothermal fluid may have combined, releasing hydrogen and making iron pyrite. Commonly derided as "fool's gold," pyrite may have helped create the most valued of riches—life itself. A snippet of pyrite may have been the home where carbon was first assembled into the building blocks of life.

If deep-sea hot springs were the beginning, the dawn of life occurred in a dark inferno. Near the roots on the tree of life are ancient heat-loving bacteria, Archaea, whose descendants still thrive in the sea's scalding hydrothermal vents. Medical instruments typically autoclaved at 250 degrees Fahrenheit (121°C) are sterilized, killing all bacteria known to humans. Strain 121, a bacterium recently plucked from the fluid of a North Pacific black smoker named Finn, is an exception. After twenty-four hours in this hellishly hot water, Strain 121 doubles in number, its tenacity restructuring our understanding of the kind of extreme environments that harbor life.

Whether the birth of earth's first living organism was an accidental and unique occurrence that happened just once, or whether it was a natural, perhaps inevitable result of the right ingredients being in the right place at the right time, we have yet to discern. We have yet to know whether the spark of life was lit just once or many times. Although our eyes into the past sharpen, the image remains hazy.

Vestiges

Strain 121 requires light for growth. Other ancient heat-loving bacteria abounding in the scalding vent water live on sulfur. During an eruption, they are jettisoned from the vent chimneys, creating a blinding blizzard of bacteria swirling 160 feet (50 m) above the seafloor. Fossil remnants of their forebears, though, are rare, and vestiges of earth's first life hard to identify. Few have survived three and a half billion years of earth's history, and those that have are barely recognizable.

The outback of western Australia's Pilbara Craton holds some of earth's oldest rocks. Beneath a remote landscape of red rock and rust, broken by the occasional acacia and eucalyptus tree, lie the remains of an ancient deep-sea hot spring more than three billion years old. Encased in a massive deposit of sulfide are tiny threads of pyrite, possible Archaean fossils from this early time. The spring originated in deep water about 3,300 feet (1,000 m) below the surface, offering shelter from the sun's deadly ultraviolet rays. The water in this early sea held little or no oxygen, and no sunlight penetrated this depth to allow photosynthesis; the single-celled organisms that dwelled here probably fueled their metabolisms with sulfur in temperatures that most likely reached at least 212 degrees Fahrenheit (100°C).

If these thin, sinuous filaments prove to be true fossils, they are the oldest surviving from a deep-sea hot spring. The ocean is the cradle of life. Its basin holds the early history of evolution. That history is told by organisms that came to rest in the sea, where they turned to rock that later became dry land. Other even older, possibly fossil-bearing rock dating from this early time lies north of the remnant hot spring. These ancient rocks, 3.4 billion years old, formed in shallow waters where the tide has long since ebbed; at the site today, water flows seasonally when ribbons of deep sand turn to broad rivers during the monsoons.

Across the Indian Ocean lies additional evidence of earth's early sea and the life it held. Glassy pillow-shaped lavas erupted from an ancient seafloor lie exposed along the banks of the Komati River, in South Africa's rugged Barberton Mountains.

Carbon is the foundation of life. It exists in every living organism, in every cell. While some is stilled, preserved in fossils over long stretches of time, most is continually recycled. A cell dies and decays, and its carbon atoms are released and reused to support another cell in another organism. Humans are mostly water and, after that, mostly carbon—carbon that has been passed down through the ages, from the flesh of a fish, the ear of an elephant, the leaves of a plant. Somewhere in each of us is a cell whose carbon elements may have nourished the planet's nascent life.

We hold the traces of our ancestry in other ways as well. Proteins, essential to the functioning of every cell, contain nickel, copper, and sulfur—remnants of their early life in the metal-rich waters of deep-sea hot springs.

Some of earth's oldest life is preserved in the rocks of Karijini National Park, Pilbara, Australia. Opposite, gorges, and below, an iron plant (Astrotricha hamptonii).

See Map 1, Fossil Sites, p. 264.

Endurance

Early life was astonishingly versatile. Our metabolism runs on oxygen; our bodies consume, rather than produce, food. Earth's early microbes produced their own food and consumed what others produced. Their metabolisms ran on iron, sulfur, methane, and even arsenic. In a sea without oxygen, they used energy from hydrogen plentiful at the vents. They lived in the boiling, acidic waters of deep-sea hot springs, in conditions where, until recently, we could not imagine any life.

They have endured. Today thousands of strains live in the sea's hydrothermal vents, in the corrosive and toxic wastewater from mining operations, and in the Middle East's extraordinarily salty Dead Sea and Utah's Great Salt Lake. Constantly swapping genes, adapting to the most hostile of environments, they are a storehouse for the myriad possibilities of existence, and the foundation from which all other life evolved.

How they came to harness the light of the sun to power their metabolism is a mystery still unfolding. GSB1, the bacteria living at Nine North that use energy from dim vent light to manufacture food, may provide a link: plumes of water pouring from the vents may have carried GSB1's ancestors into the shallow ocean, where their photosynthetic capability adapted to sunlight.

Right: Ancient strains of bacteria endure in extraordinarily salty Lake Assal, Djibouti, in the low-lying Afar Depression, where a new ocean is being born.

Opposite: A computer-generated image of a dividing cell: microbes were the first life to appear on the young earth. From them all life evolved.

A Life-Changing Force

The ancient Archaean world extended from 3.9 billion to 2.5 billion years ago. During that time, tectonic forces created earth's first sea. Enriched by mineral-bearing vent water, this primordial sea spawned and nourished earth's first life. While physical and chemical forces created earth's first living cells, those cells would grow and prosper to become a mighty force of their own. In time, life would change the face of its creator.

Bohar snapper gather to spawn in the Red Sea. The sea sustains all life, whether it dwells on land or in the water.

Touched by the Sea

2 Building the Basin

Wave breaks over coral reef, Ailuk Atoll, Marshall Islands, Micronesia. The Pacific Ocean is closing.
Eventually Micronesia, along with other Pacific islands, will become part of the continents.

Odd as it may seem, it would not be remiss for beachcombers to find traces of the sea inland, away from shore. At the foothills of the Italian Alps, not far from Verona, the limestone-layered Pesciara (fishbowl) contains fossils of jellyfish, mackerel, and shark so well preserved that even their internal organs are outlined. High in the Himalaya are coiled shells of ancient marine mollusks. In Canada's Rocky Mountains, layers of rock hold many strange and beautiful fossils of earth's earliest animals. They too dwelled in a sea long gone.

So many pieces of seafloor, so far from water, defy the very concept of solid land. Charles Darwin, hiking through the Andes and finding seashells up at the Continental Divide, was awed. "Daily it is forced home on the mind of the geologist," he wrote, "that nothing, not even the wind that blows, is so unstable as the level of the crust of this earth." From the sea comes land we call home.

Seafloor rising into the mountains is born at mid-ocean ridges, where hot lava doused by cold water builds ocean basins. Some are older than others. The Atlantic is a young, growing sea, widening little by little, year by year. As it grows, Europe and America drift slowly apart. Older seas, the Pacific and the Mediterranean, are aging, their basins narrowing. As they close, continents collide. Mediterranean seafloor sinks back into the earth, and Africa lurches toward Europe.

Ever since earth's early days, ocean basins have come and gone, opening and closing again and again. The distance between young seafloor at mid-ocean ridges and old seafloor at the edge of continents is great, and the water—earth's largest dwelling space—is deep.

The mountains of the Mid-Atlantic Ridge, where new seafloor is formed, rise above the waves at Ascension Island in the South Atlantic (above) and at Surtsey near Iceland (right and far right).

The Floor of the Sea:
From the Mountains to the Plains

"Could the waters of the Atlantic be drawn off, so as to expose to view this great sea-gash, which separates continents and extends from the Arctic to the Antarctic, it would present a scene the most rugged, grand, and imposing. The very ribs of the solid Earth, with the foundations of the sea, would be brought to light." Matthew Fontaine Maury, U.S. naval officer and oceanographer, wrote those words in 1855, when sailors plumbed the depths of the sea with cannonballs tied to bailing twine. Little did he know how rugged, grand, and imposing modern science would show the seafloor to be.

Maps of the sea made with sound waves, and thousands of cores taken from deep within the seabed, reveal mountains more rugged than the Andes or the Himalaya, cut by steep valleys that rival the Grand Canyon. The once invisible seafloor, not the land resting in plain sight, is earth's paramount feature.

The youngest, newest seafloor lies in the craggy mountains of the ridges, in the steep valleys where lava wells up from the depths. The mountains tower above the seabed and occasionally above the waves, on Iceland and on Ascension, a lone volcanic island in the South Atlantic. Pieces of volcanic rock, already split as the seafloor moves apart, face each other across the valley. The symmetry startles.

Seafloor widens slowly but not smoothly, and the mountains are scarred by their violent genesis. The valleys are strewn with piles of rock shattered by earthquakes. New seafloor is added at different times in different places along the ridge: the moving segments scrape by each other, opening deep and jagged chasms between them.

Time smoothes these rough edges. New mountains continually emerge, yet the ridges are but narrow seams winding through the ocean. As the cracked plates of earth's surface drift apart, mountains slide away from the place of their birth. As they age, they cool and subside into the seabed, losing height and grandeur. Erosion rounds their peaks, and particles drifting from the surface rain down upon them. Dust blown in on the wind and washed to the sea in rivers, and tiny shells from single-celled plants and animals once drifting at the sea surface, bury the mighty mountains. Ever so slowly, the mountains turn to hills, and the hills disappear into the wide, flat plains of the deep abyss.

The abyssal plain, flatter than the prairies of the Great Plains, stretches from the foothills of the ridge across the sea to steep slopes rising to the continents. Oceanic mountains and plains are vast; the basins of the sea comprise fully two-thirds of the planet's surface. The water above these features—deep, dark, and bitter cold—is, by volume, 90 percent of the sea.

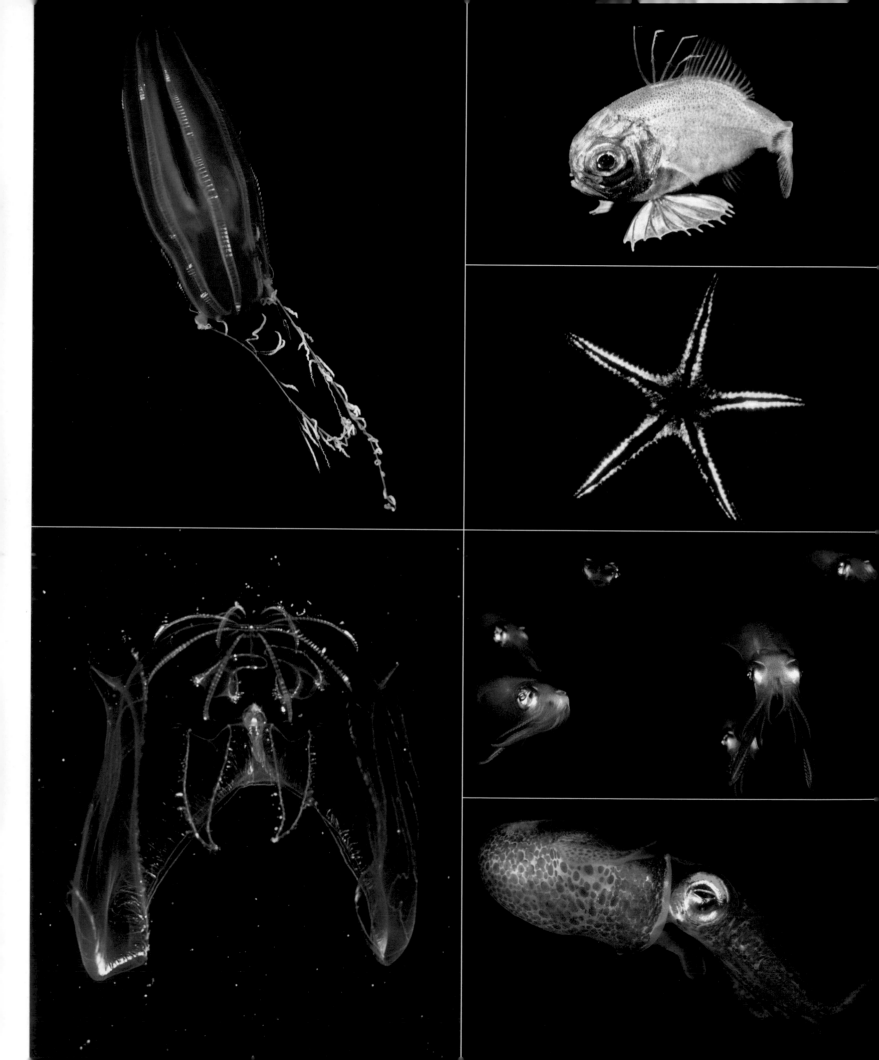

While many dwellers of the deep sea produce light themselves, others, like the Hawaiian bobtail squid, *Euprymna scolopes*, acquire their light producers, trapping bioluminescing bacteria from the water. The bacteria, *Vibrio fischeri*, communicate, sensing their numbers increasing in the squid's photophore. When they reach a quorum, they begin to generate light. Their relationship with the squid is symbiotic. In return for sustenance, they enable their hosts to survive in a treacherous world.

Bioluminescence in the deep sea has evolved many different times and across many species. Left top: Black-gut red cydippid ctenophore, West Atlantic. Right top: Two-fin flashlight fish (Anomalops katoptron), pelagic stage larvae, Indonesia. Right middle, top: Sea star (Plutonaster bifrons), Northeast Atlantic. Right middle, below: Bigfin reef squid (Sepioteuthis lessoniana), Indonesia. Right bottom: Common bobtail squid (Sepietta oweniana), Norway, North Atlantic. Left bottom: Ctenophore (Bathocyroe fosteri).

One squid's lifeline is another creature's poison. Other marine vibrios, such as *Vibrio cholerae*, communicate with each other to sicken their hosts. The realization that marine bacteria speak to each other has opened whole new avenues of research as scientists seek antibiotics that could potentially block the communication.

The soft, black-bodied vampire squid, *Vampyroteuthis infernalis*, confuses its attackers by folding its webbed arms over its mantle. As it darts away from a predator, it squirts not ink but a viscous bioluminescent fluid, wrapping itself in a glowing cloud of light. A beautiful comb jelly from the Gulf of Maine, *Bathocyroe fosteri*, is transparent except for its red stomach, which hides any bioluminescent animals the jelly may swallow, preventing its dinner from luring predators.

Wonders Unfolding

So little of the deep sea has been explored that almost every trip yields animals previously unseen by humans. The population of the deep sea is low in numbers but high in diversity. The presence of animals in an environment so remote from our own continually expands and redefines the possibilities of existence. The deep sea is a sea of never-ending surprise.

We had assumed that most bioluminescent animals of the sea emit green or blue-green light, but scientists have now identified dragonfish that hunt with flashing red lights, and a colonial siphonophore that snares its prey with hundreds of red lures positioned near the stinging cells of its tentacles, asking us to reconsider the role color may play in the ecology of the dark sea.

Observations of squid in their own habitat reveal an unexpected tenderness. Most squid leave their eggs on the seafloor, but the brooding squid, *Gonatus onyx*, normally a shallow-water resident, remains in deep water after spawning, holding its large and cumbersome egg sac in its arms. There it remains, living off stores of lipid and rocking the sac to aerate it. Between six and nine months later, the eggs hatch, two to three thousand of them. The parent, exhausted and weakened from standing watch so long, dies, while the next generation floats to the surface, beginning life anew.

Despite their enormous size, we rarely see the sea's largest invertebrates, the giant (*Architeuthis dux*) and colossal (*Mesonychoteuthis hamiltoni*) squids, alive. Only rarely have they been found or photographed in the deep, dark water where they dwell. We hunt their prey—the Patagonian toothfish (*Dissostichus eleginoides*)—so occasionally the squid are caught in the nets of deep-sea fishing trawlers. We infer their existence from their chitinous and undigestible beaks that show up in the stomachs of their

Sperm whales are one of the major predators of the giant squid.

predators, sleeper sharks and sperm whales. With eyes as big as dinner plates and a body the length of a bus, these squid are enormous, but their lives are obscure, testament to the still vast and unknown world of the deep sea.

To date, some of the largest known populations of deep-sea dwellers are tiny, microscopic organisms, making their home not in the water itself, but rather in sediments below the seabed—the layers upon layers of sand, silt, and shell, millions of years old and hundreds of feet thick, that bury the mountains of ridges as they age.

Scientists have found unexpectedly large numbers of living organisms hidden in these buried sediments. Single cells, they may account for 30 percent of life on earth. Scientists drilling in the seabed of the Pacific Ocean off the coast of Peru found them in astonishing numbers—between one billion and a hundred billion organisms in half a teaspoon of sediment.

Indigenous to the seabed, they are unrelated to previously known, named lineages. Their very existence defies our ideas of what constitutes life. These microbes are alive, but barely. They contain ribosomes that signal the presence of a living cell, but they grow ever so slowly, replacing themselves, on average, once every thousand years. Scientists have yet to determine how they manage to survive on such slow metabolisms. Their lives, like the lives of the giant squid, are still beyond our grasp.

From the Plains of the Abyss to the Dry Land of Continents

Seafloor freshly built in the mountains of mid-ocean ridges cools and subsides. Shells of tiny snails—pteropods—and single-celled coccolithophores drift from the surface, burying the mountains in the endless flat plains of the abyss. In the deep water, carbon dioxide dissolves the shells. Then fine dust blown in from land builds the seafloor into red clay that accumulates ever so slowly, a thin layer every one thousand years. Solitary deep-sea spiders, brittle stars, and sea cucumbers make their way along the bottom; trails in the mud suggest who may have passed by before.

By the time young seafloor has moved from the place of its birth across the wide abyssal plain to reach the edge of continents, half a mile to nearly 2 miles (1–3 km) of clay and dust have accumulated. Although millions of years old, seafloor is young. Cores of continents endure through eons of time—much of Canada, over two billion years old, has witnessed more than half the planet's history—while ocean basins make but a brief sojourn. Two-thirds of earth's surface is seafloor, a mere 200 million years old—a geologic infant whose days amount to only 4 percent of earth's history. What is born at mid-ocean ridges will, for the most part, return to the depths. It does not go gently.

At 9:40 on the morning of November 1, 1755, All Saints' Day, churches and other buildings began shaking, then collapsing, as an earthquake struck the city of Lisbon. The earthquake, which occurred off Portugal's coast, generated a tsunami that capsized and destroyed boats and cargo lying in Lisbon's harbor. The giant wave then flooded the city, drowning those who had rushed to the river for safety. By afternoon, one of Europe's wealthiest cities had been reduced to rubble.

Volcanoes erupt violently as seafloor born at mid-ocean ridges descends back into the earth at the edge of continents. Bottom left: Mayon Volcano, Philippines. Bottom center: Fuego Volcano, Guatemala. Bottom right: Arenal Volcano, Costa Rica.

There were those in Lisbon who feared that an angry God had punished his sinning people, but the destruction originated beneath the waves in the motions of a restless earth. To human beings going about their daily lives, that motion is often slow, imperceptible. Earth-shattering events like the 1755 Lisbon earthquake, the 2004 earthquake and tsunami in Indonesia, and the eruption of the Philippines' Mount Pinatubo are violent manifestations of a planet renewing and rebuilding itself. Lisbon was destroyed as seafloor, beginning its descent, buckled.

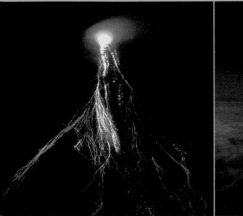

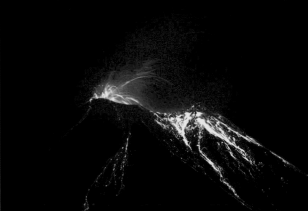

See Map 4, Earthquakes, p. 267; and Map 5, Volcanoes, p. 267.

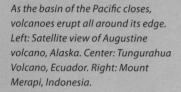

When the plates that crack earth's surface pull apart at mid-ocean ridges, new seafloor is born. When they collide, seafloor begins its descent back into the earth. The return takes place in the deepest parts of the ocean, close to shore, in trenches—huge gashes that run along the Pacific edge of South America and along the archipelagoes of Indonesia, the Philippines, and the Aleutians. We cannot see the descent taking place below the waves, but we feel it. Land rimming the Pacific, the Ring of Fire, is restless, prone to earthquakes, tsunamis, and violent volcanic eruptions.

As the basin of the Pacific closes, volcanoes erupt all around its edge. Left: Satellite view of Augustine volcano, Alaska. Center: Tungurahua Volcano, Ecuador. Right: Mount Merapi, Indonesia.

At the western rim of the Pacific, beneath the islands of Indonesia in the heart of the Ring of Fire, two pieces of ocean floor collide at a deep-water trench. On December 26, 2004, an earthquake at the site created a tsunami. Within minutes an 82-foot (25-m) wave rushed onto Sumatra, destroying everything in its path and killing 200,000 people. Within two hours, the wave had crossed the ocean to Sri Lanka and Thailand, killing 65,000 more. The earthquake propagated through 800 miles (1,300 km) of seafloor and jolted one plate 33–65 feet (10–20 m) toward the trench, altering earth's gravitational pull and changing the path of satellites.

The seafloor continued its descent. Three months later, on March 28, 2005, it shifted again, this time tilting the island of Nias, off the coast of Sumatra, into the sea. It permanently flooded one side of the island, washing an entire village into the water, and raised the other side, draining the harbor and beaching mangroves and coral reefs. Yet another earthquake took place in July 2006, killing 600 people. More are to come: the motion of the sea is inexorable.

Land Redesigned by a Closing Ocean Basin

From these convulsions, continents grow: from layers of rock scraped off descending seafloor and pushed into rising mountains such as the Himalaya and the Andes; from seafloor melting during its descent and creating explosive volcanoes like Mount St. Helens; from arcs of islands—Indonesia, Japan, the Antilles—created by colliding seafloor, then plastered against the continents. Most of earth's 500 to 600 active volcanoes are born as seafloor is created or destroyed. Colombia and Venezuela, Panama and Costa Rica, New England and Nova Scotia: all contain remnants of ancient island arcs that once belonged to the sea.

See Map 2, Plate Tectonics, p. 266.

Steep, flakey, gray cliffs at the water's edge in Newfoundland's Gros Morne National Park are upended pieces of ancient seafloor. Further into the park lies a barren, orange landscape where little grows. It too comes from the sea. The rocks of the Tablelands originated deep within the earth, in the mantle below the seabed. Millions of years ago, as an ancient ocean, Iapetus, closed, they were lifted onto Newfoundland, providing a rare view of earth's deep interior. In the Troodos Mountains of Cyprus are rich veins of copper, originally deposited at hydrothermal vents in an ancient sea. Cut into the dry landscape of Oman are ancient hydrothermal vents and pillow lavas that a departing sea bequeathed to land. The rich soils of America's Great Plains were laid down in a warm, shallow sea.

The ancient floor of the sea is written into the landscape of continents and has been since continents began to grow, early in earth's history. Folded into a small rocky hill in an isolated cove on the coast of Greenland, is a chunk of earth's oldest rock. It contains a piece of seafloor formed at an ocean ridge, carried across a sea, and then uplifted onto land, where it has sat unscathed for 3.8 billion years.

Rock this old, among the first lines in earth's biography, is rare. More often, as soon as seafloor is raised into mountains, wind and rain wash them back into the sea, where their minerals will nourish the sea's inhabitants over and over again, before turning to shell or stone, building another range of lofty mountains that will again erode into the sea. An opening and closing ocean is earth recycling writ large. With each turn, there is new life.

Just as ancient seafloor is built into today's continents, when the young Red Sea closes, it too may become part of the land.

Breathing

Oxygen was deadly for earth's first inhabitants. Highly volatile, altering what it touches, it would have incinerated the life taking hold on the young planet. Eventually oxygen would fill earth's early sea, making our lives possible, but in the beginning this vital gas, now essential for our own existence, was rare. Tiny bacteria made the difference.

Earth's Young Sea: Iron-Rich

Earth's youthful ocean was rich in dissolved iron, brought up from the planet's molten core and released through hydrothermal vents. Rocky outcrops lining the Canadian shore of Lake Superior contain remnants of that ancient sea. Two billion years old, the Gunflint chert lies at the edge of the lake against a backdrop of balsam forest. Chert, extraordinarily hard, can spark a fire, be chipped into an arrowhead, and ignite gunpowder. Resistant to the ravages of time, it keeps earth's history.

Background: Two-billion-year-old rocks, lining the shore of Lake Superior, were once part of an ancient sea.

Iron-loving bacteria thrived in earth's early sea, and their fossil remains are locked within the Gunflint chert. The sea is long gone, but a living likeness exists high in the Arctic in the Gallionella Garden hydrothermal vent field, where orange-yellow iron-loving bacteria carpet the seafloor. The Gunflint chert records the iron-rich character of earth's early ocean. In the Gallionella Garden, earth's past lives.

A Whiff of Fresh Air

For hundreds of millions of years, life on earth was powered by hydrogen released at deep-sea hot springs. A more abundant and widespread source of hydrogen existed, untapped. Water covered the planet; each molecule contained two atoms of hydrogen along with one of oxygen. Under circumstances we have yet to fully discern, in a moment we have yet to fully characterize, a microbe split a water molecule, seized the hydrogen for its own use, and released the oxygen as waste. It was a momentous act. The impetus for this critical stage in our evolution remains a deep mystery still to be unveiled.

Perhaps a shortage of usable hydrogen fueled this innovation, favoring those microbes that, through gene swapping or natural selection, evolved to capture the hydrogen so plentiful in the sea. Whatever the circumstances, bacteria were freed from the confines of hydrothermal vents. A photosynthetic wonder resulted. Tapping into a

source of energy far greater than anything the vents provided, bacteria powered by water and sunlight spread and multiplied and came to dominate the sea. Deep-sea hot springs may have spawned life on earth, but the progeny of those early cells ultimately traveled far from the place of their birth, developing lives of their own and succeeding in ways impossible for their ancestors.

Oxygen changed everything. Crashing meteors and seething volcanoes shaped earth's early atmosphere, creating the setting where life evolved, but now life itself became an equally powerful sculptor of the planet. Tiny, single-celled organisms, invisible to the naked eye, filled the air with their waste, oxygen, bringing about a transformation in the atmosphere that hasn't been seen since. Never was such a widespread pollutant put to such good use. Single cells of chlorophyll floating in the sea radically altered earth's atmosphere, resetting the terms for existence and creating an environment that would enable the rise of animals.

Those That Imbued the Atmosphere with Oxygen

Their time on earth is remembered by the sea, which records so much of evolution's story. Embedded in 2.7-billion-year-old pieces of seafloor, stacked layer upon layer to build the bright red gorges of western Australia's Karijini National Park, are tiny signatures of these organisms—traces of lipids from membranes that held the cells that revolutionized respiration. These cells were ubiquitous: their signatures are also present in rock that once rested in the delta of an ancient sea but today lines the shore of Ontario's Elliot Lake, deep in moose country.

Above: The oxygen that filled earth's atmosphere was first generated by single cells of chlorophyll floating in the ocean.

Right and far right: Their remnants line the gorges of Australia's Karijini National Park.

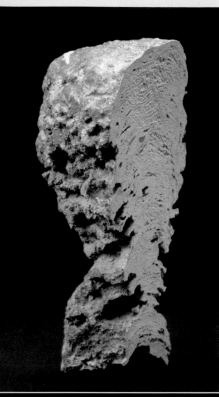

Once they were called Cryptozoa, "hidden life." Not only were the bacteria themselves hidden within the rock, their very nature was hidden from us as well. These single cells of chlorophyll, known today as cyanobacteria for their blue-green tinge, redesigned earth's atmosphere, breath by breath. Earth's neighbors, Venus and Mars, testify to their work. Their atmospheres, like that of the young earth, still contain little oxygen, while today earth's atmosphere is fully 21 percent oxygen, pumped in by living cells.

They dwelled in warm, shallow water. An early example of communal living, they lived together in thin, sticky mats that trapped bits of sand and grit. The colony moved ever upward, toward the light, as the accumulating layers turned to stone. These dome-shaped stromatolites, "layers of stone," memorialize the bacteria that filled the atmosphere with oxygen. Stromatolites once formed great reefs at the sea's edge. Today their remains are scattered throughout the world: in the snowy Belcher Islands of Hudson Bay, the icy Arctic archipelago of Svalbard, the desert rocks of Australia, and the cliffs above the Kotuikan River in Siberia.

The days when stromatolites dominated the sea have come and gone. Breathing oxygen into the atmosphere, they created a world where their role would diminish. Their gift, an infusion of oxygen, gave rise to animals that grazed them away. Considerably less widespread today, they live on in the salty lagoons of Baja California, Mexico; in steep channels among the Bahamas, where the current runs swift, scouring away other forms of life; and in the far reaches of Shark Bay, Australia, in a warm, shallow pool behind a sandbar. This salty pool, beyond the beaches where sea turtles come to nest, beyond the sea-grass meadows where sea mammals come to feed, is crowded with clumps of stromatolites. In water otherwise inhospitable to life, their stony mounds are a living memorial to tiny sea dwellers that first gave us air to breathe.

The bacteria themselves also persist. After 2.7 billion years, these single cells of chlorophyll floating at the sea surface continue to flourish virtually unchanged, retaining their essential identity generation after generation. They survive by means of their genetic versatility. Constantly swapping genes, some introduced through viral "infection," they easily adapt to changes in water depth, temperature, and availability of sunlight.

In today's ocean, modern descendants of these bacteria, *Prochlorococcus* and *Synechococcus*, are the smallest known marine microbes. Each a hundredth of the width of a strand of human hair, they are minimalists with the fewest genes of any organism that captures energy from sunlight. Small in size but great in number, they are the most abundant photosynthesizing organisms in the sea. Hundreds of thousands of them fit in a spoonful of seawater.

Prochlorococcus and *Synechococcus*, along with the other photosynthesizing organisms of the sea, are critical to our well-being, carrying out almost half of earth's photosynthesis, producing half our oxygen. We can't live without them. A long time ago, they created an atmosphere that we could breathe, and now they sustain us while we are here.

A Partnership

The sea's photosynthesizing bacteria did not create earth's new atmosphere alone. Emerging life and land had entered into a fertile partnership that helped give earth its oxygenated atmosphere. Today hard granite—exposed in the peaks of the Himalaya and the gorges of the Grand Canyon, underlying African savannas and English moors, built into the stone walls of New England and the obelisks at Karnak—is the foundation of earth's continents. It is ubiquitous on dry land but rare elsewhere. The surfaces of the moon, Venus, and Mars are mostly made of heavy basalt, the rock of ocean basins. Early in earth's history, photosynthesizing bacteria in the sea helped make granite continents.

Stromatolites live on today (in Shark Bay, Australia, above), and their fossils (opposite) are found through-out the world.

Granite forms the foundation of continents. Below: Granite outcrop in the Boundary Waters Canoe Area Wilderness, North Woods, Minnesota.

Bacteria turned energy from the sun into chemical energy that weathered and altered the rock of the sea. That seafloor collided with other seafloor or small basalt islands (like Hawaii or Iceland) that dotted earth's early ocean. During its descent back into the earth, the altered rock, containing more water, melted at a lower temperature, producing the lighter granite that constitutes earth's continents. As young seafloor born at ancient mid-ocean ridges drifted, became altered, and returned to the depths, it produced more granite, and the continents grew larger and more stable. They in turn would enable the atmosphere to hold oxygen produced by marine bacteria.

Before there were many continents, earth's volcanoes erupted in the sea, spewing hydrogen and hydrogen sulfide that consumed oxygen produced by marine bacteria. As continents grew, the number of continental volcanoes increased. These volcanoes, erupting at higher temperatures, emitted gas that didn't react with oxygen, allowing it to accumulate in earth's atmosphere.

For much of earth's early history, the sea lacked oxygen. For many sulfur- and methane-eating organisms that thrived in that sea, the arrival of oxygen was a catastrophe. They retreated, seeking refuge from the polluting toxin, and today they linger on in waters where little or no oxygen is found: in thick mud and swamps, in highly salty seas and the fluids of hydrothermal vents, in crevices deep below the seafloor. However hidden, they remain essential to life on earth, cycling and recycling sulfur that helps build key amino acids and proteins in plant and animal cells. They hold among them many blueprints for existence, many methods to capture energy and metabolize food. They have endured for 3.5 billion years of earth's history, and we for a mere 200 thousand years. They have survived one environmental catastrophe after another, one mass extinction after another. Should circumstances change, should the conditions for life on earth be reset, in all likelihood it is these tiny organisms born of the sea—not we—who will survive.

The Sea Rusted

Oxygen made its mark in the sea, but not immediately. Earth's primordial ocean remained rich in iron spewed from the hydrothermal vents until growing numbers of marine bacteria puffed oxygen into the sea and changed its chemistry. Touched by oxygen, the dissolved iron precipitated and fell to the seafloor. Eons later, some became dry land holding the world's richest iron deposits.

In the United States, those iron deposits are in the Marquette Iron Range of Michigan's Upper Peninsula. Iron from Michigan, one of the United States' leading iron producers, fueled the Industrial Revolution. Many of the mines are depleted, while others continue to operate. The legacy of iron from earth's early ocean is there to see: the exposed rock of Jasper Knob, a bald-topped hill in the town of Ishpeming, is striped with alternating bands of iron and jasper, recording a time when marine organisms breathed out oxygen and rusted the sea.

The world's richest iron deposits were formed in an ancient ocean. From the Marquette Iron Range, banded iron (above) from Jasper Knob, Ishpeming, Michigan, and water reddened by tailings from an Ishpeming iron mine (right).

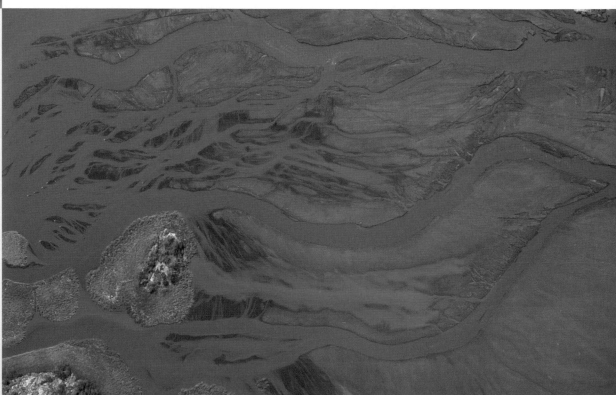

Intimation of Life to Come

The rocks of the Gunflint chert strewn along the shores of Lake Superior hail from the ocean of earth's youth. Other fossils intimate the sea and the life that would come. At some time in earth's history—no one is certain when—single cells of life joined, pooling resources. Perhaps one captured the other, perhaps they embraced. A cell that combined motion and metabolism resulted. The new cell, carrying its DNA within an enclosed membrane, would contain energy-producing mitochondria and a chloroplast that resembled the photosynthesizing cells of floating marine bacteria. Single, individual, independent cells came to live together, communally and then symbiotically, each needing the other. The whole, a multicellular energy powerhouse, would come to be far greater than the sum of its parts.

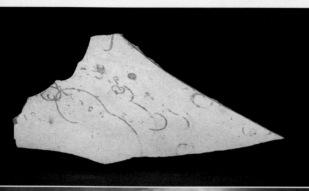

The desert of western Australia is rich in fossils that record the early evolution of multicellular organisms. The earliest record is but a tantalizing hint: 2.7-billion-year-old rock that holds molecular fingerprints, smudges of sterols (of which cholesterol is one) that once strengthened the membranes of earth's first nucleated cells. Oil trapped in ancient rock at Ontario's Elliot Lake holds these signatures as well.

Farther east, in the Roper Valley in northern Australia, layers of shale laid down in a vanished sea reveal other signs: Top End holds some of the bottom layers of evolution. Beneath the tick-infested scrub are rocks containing tiny spherical cells, one and a half billion years old, that seem to have changed shape during their lifetimes, a key signature of a multicellular organism. Back in the Upper Peninsula of Michigan, thin coiled strands are imprinted in the rock of the iron mountains—fossils of red algae, *Grypania*, from the dawn of multicellular life.

Top: This Grypania spiralis *fossil in the Marquette Range, Michigan, dates from the dawn of multicellular life.*

Above: Ireland's Connemara coast: on rocky intertidal zones throughout the world live red seaweeds descended from earth's first multicellular organisms.

Farther north, on Somerset Island in the Canadian arctic, is yet another sign. The Hudson's Bay Company once ran a trading post on the island, a way station along the Northwest Passage. Long before the English arrived, Thule ancestors of the Inuit hunted bowhead whales from its shores. Even earlier, 1.4 billion years ago, Somerset Island was part of a shallow sea. A red seaweed, *Bangiomorpha pubescens*, grew in its waters, anchored to the bottom with a holdfast. The presence of the holdfast, as well as spores and gametes, indicate clear differences in purpose among the organism's cells. Unlike the sea's single-celled life, *Bangiomorpha* reproduced sexually, opening a critical route for the exchange of genetic information and for the diversification of species.

The icy rocks of Somerset Island hold the memory of this organism, earth's first known multicellular, sexually reproducing life. *Bangiomorpha* has endured. Its descendants, *Bangia*, virtually indistinguishable from the red algae sealed in the ancient rock, live on rocky pilings in Chesapeake Bay's York estuary and on the stony shores of Galway, Ireland, and Marblehead, Massachusetts.

Earth's Barren Middle Ages

Oxygen is a powerful energy source, four times more powerful than other metabolic pathways that sustain life. The ability to breathe oxygen would eventually give rise to an explosion of life in the plant and animal kingdoms, enabling a single cell of bacteria to evolve into many large organisms. The seeds sown by the nameless nucleated cells in Australia, by *Grypania*, and by *Bangiomorpha* were not reaped for hundreds of thousands of years. The lag time between the oxygenation of the atmosphere and the rise of multicellular animals was two billion years. The blueprint of life as we know it evolved in the ocean, but the sea was slow to fill with this life-giving force.

Frozen rocks along the Black Sea coast near Tsarevo, Bulgaria. The deep waters of the Black Sea today are, like earth's early ocean, devoid of oxygen.

While the atmosphere filled with oxygen, the diversification of ocean life remained stunted, its evolution on hold. While the sea had given birth to and sustained earth's early life and would nourish the life that would follow, during this middle period of earth's history it couldn't. Ancient pieces of seafloor in Barney Creek, Australia, suggest why.

Layer upon layer of rock, resting undisturbed in the lush mountains of northern Australia west of the Gulf of Carpentaria, in the aboriginal lands in and around Kakadu National Park, hold lipid signatures of microbes that lived in the sea 1.6 billion years ago. The atmosphere then was suffused with oxygen, but the sea, according to the story recorded within the rock, was not. Droplets of oil squeezed from these rocks and analyzed for molecular markers describe a sea filled with green and purple bacteria that lived without oxygen, deriving their energy from sunlight and sulfur.

The deep waters of the Black Sea, like earth's early ocean, are devoid of oxygen. Not far from shore, trading ships hundreds of years old are preserved in the anoxic water, their wooden hulls safe from the voracious appetites of wood-boring animals. The oxygenless sea protects the ships and their cargo and, with them, an important history of maritime trade.

No longer rich in dissolved iron as it had been—but not yet rich in oxygen as it would become—the sea of earth's middle age was filled with sulfur. Weathered from land and washed into the ocean, it took a toxic form, hydrogen sulfide, that would poison the eukaryotic life (cells with a nucleus) that was emerging. Furthermore, this sea lacked the nourishment necessary for life that would soon burgeon in its waters. Hydrogen sulfide removed iron, molybdenum, copper, zinc, and other trace metals—essential nutrients for larger forms of life. Evolution stalled because the sea was malnourished.

In time the sea would replenish itself, supplied with nutrients when continents fused and an ancient ocean basin closed. Along the eastern seaboard of North America, from Newfoundland to North Carolina, ancient seafloor was pushed high into the Adirondacks, whose mountains were once as tall as the Himalaya are today. As soon as the mountains rose, they began to erode. Wind and rain disintegrated their rock, washing minerals and metals into the sea. These minerals and metals, combined with continued increases in oxygen, made it possible for marine organisms, whose evolution had proceeded ever so slowly, to obtain the nitrogen necessary to fuel their growth. Feeding its inhabitants well, the newly oxygenated sea gave rise to a startling explosion of life.

3 Circle of Water

Off the coasts of Iceland (shown) and Greenland, cold water sinks to form the North Atlantic's deep currents.

At the edge of the Arctic Circle, between Iceland and Greenland, lies a narrow, icy channel, the Denmark Strait. Here, hidden beneath the water's cold, gray-blue surface, unseen by human eyes, is an unexpected surprise: earth's most spectacular waterfall. Every second, nearly 106 million cubic feet (3 million m^3) of water spill into the North Atlantic over an undersea sill of volcanic rock connecting Greenland to Iceland. The world's better-known waterfalls—Victoria Falls, 35,000 cubic feet (1,000 m^3) per second, or Niagara Falls, a little more than twice that—are mere trickles by comparison. The cataract in the Denmark Strait is on another scale altogether.

No one comes to admire the grandeur of this cascade. No one can. It is described not by people but by an array of current meters resilient enough to withstand corrosion and the rush of water. Its name does it little justice: the Denmark Strait Overflow is that, yet so much more. Like earth's terrestrial waterfalls, it feeds a river—an undersea river that cascades over the undersea sill and flows through the ocean, mightier than any river cutting through dry land. The waterfall, the river, and its tributaries are all joined in an endless flow of currents circling the earth. Just as a closing ocean basin creates land where humans dwell, and just as the sea's tiny single-celled organisms breathe oxygen into the atmosphere for us to breathe, so too does this coursing water sustain us. Currents bring oxygen to the deep water and carry the sun's heat to icy latitudes. They shape the wind, giving moisture that soaks continents with rain. On them our lives depend.

Ubiquitous Sea

The surface of the earth is mostly water, and that water is mostly sea. The ocean covers more than twice as much of the earth as dry land, but this common description understates the vastness of the sea, so much deeper than the land is high. Mount Everest, earth's tallest peak, could easily fit within the deepest water, the Pacific's Marianas Trench, but that is one mountain in one ocean basin. All earth's dry land could easily be hidden within the sea. Land rising above the sea is only a fraction, less than 10 percent, of the water below. The water realm of our planet is immense, and dry land, by comparison, minuscule; sea, not land, is earth's distinguishing feature.

Though the sea endures, its aspect is inconstant. Over four billion years of planetary history, as continents have come together and broken apart, ocean basins have come and gone, widening and narrowing, flooding the land and then receding. Today oceanographers describe five ocean basins: the Atlantic, young and narrow, growing, spanning earth's length from pole to pole, and named for Atlas, who, like the sea itself, holds the world on his shoulders; the Pacific, earth's largest, deepest, and broadest basin, named by Magellan for its calm waters, although explosive volcanoes ring its edges and typhoons batter its many islands; the Indian, whose monsoons have taken maritime traders back and forth across its warm waters for centuries, and from whose volcanic ridge was ejected the most voluminous plume of vent water ever recorded; the ice-covered Arctic, a sleeping giant whose secrets are being revealed as its ice thins; and the Southern, circling Antarctica in a ribbon of swiftly flowing water.

THE CHANGING FACE OF EARTH'S CONTINENTS AND SEA

CONFIGURATION OF CONTINENTS
LATE PROTEROZOIC PALEO (*left*)
650 MILLON YEARS AGO

FUTURE WORLD (*right*)
+250 MILLION YEARS

 Ancient Landmass

 Modern Landmass

Subduction Zone
(triangles point in the direction of subduction)

Seafloor Spreading Ridge

Earth's continents and seas have divided and joined many times. One billion years ago, earth held one sea and one continent, Rodinia, whose name means "motherland" in Russian. Rodinia's core endures in the ancient mountains of Newfoundland, the wide expanses of Canada, and the rocky outcrops of New York City's Central Park. This continent did not last. Emerging seafloor split Rodinia apart. About 250 million years ago, the pieces once again coalesced into the super-continent Pangaea, only to be cracked and flooded by the emerging Atlantic. In time, they will be made whole once again. One projection envisions that in 250 million years the Atlantic Ocean will disappear, the Indian Ocean will become an inland sea, and today's continents, now dispersed over great distances, will once again fuse, enveloped by a single ocean.

Beneath the waves at the sea surface lies enough water to easily hide all earth's continents.

Today's configuration of the sea will be gone tomorrow, its present face but a passing moment of geologic time. But no matter their shape or design, earth's ocean basins are joined into a single sea, connected by currents that circle the earth, that plunge from the sunlit surface to the distant depths, and that in time return. This journey, critical to the health of both sea and land, begins in only a few places at the far ends of the earth, where the sea freezes.

Rhythms in the Deep

In the howling winds, bitter chill, and rough waves of Arctic and Antarctic waters, seawater begins to freeze. Thin icy slivers turn to a slurry that soon hardens into thick floes. Though the sea is salty, sea ice is fresh. As seawater crystallizes, the salt stays behind, turning the water below to brine. The brine, dense and cold, sinks and spreads throughout the world. The routes of deep currents are many and varied, and the distance they travel is thousands of miles. From their place of birth in remote polar seas, deep currents fill the world ocean, accounting for 90 percent of the sea's flowing water.

Antarctic Bottom Water slides off the continental shelf and slowly sweeps north, slipping into the Atlantic, Indian, and Pacific oceans, filling their deepest basins. Most of the Pacific's deep water originates from icy brines in the sea off Antarctica.

In the Atlantic, deep water flows from the cascade at the Denmark Strait. It moves south, swelling with cold, deep water from the Labrador Sea, filling deep gouges made by shifting pieces of seafloor, its path shaped by earth's moving plates. The journey is neither straight, nor direct; the current flows along the seabed, circling through deep abyssal plains, dividing and merging, winding its way south to surface near Antarctica. There, once again, the current divides, a portion circling back into the Atlantic, the rest flowing as deep water into the Indian and Pacific Oceans.

For dwellers in the unlighted depths, the peregrinations of deep currents are a matter of life and death. The deep sea itself produces no oxygen. Giant squid and small angler-fish depend for their very existence on polar water that travels from the sunlit surface, saturated with oxygen from breaking waves and the respiration of floating plants

Tracking the Currents

Air and sea are tightly linked; pollutants wafted into the atmosphere rain out of the sky into the ocean. Chlorofluorocarbons once used in air conditioners and aerosol propellants and the fallout from nuclear weapons testing give the surface water a distinctive chemical signature, enabling scientists to follow currents disappearing into the depths. Cosmic rays hitting earth's atmosphere also strike the sea. Their rates of decay, along with oxygen levels in deep water, tell scientists when a current was last at the surface, revealing the water's age. Thousands of current meters, floats, and buoys chart the course of moving water, both on the surface and in the deep, mapping the path, temperament, and age of ocean currents. Occasionally the reverse occurs, and the currents themselves can be used to track an object lost at sea.

On the night of November 26, 1898, the steamship *Portland* left Boston on its scheduled run to Maine. It never arrived. The ship and 192 passengers and crew were lost at sea, drowned in a heavy northeaster. No one knew where the steamer went down, and those who tried to find it failed. They were looking off Cape Cod, where bodies and debris from the wreckage had washed ashore the night after the storm. It was the wrong place.

Almost a hundred years later, a physical oceanographer from Woods Hole, using wind speed and tidal data recorded on the day of the gale, plotted the path and speed of ocean surface currents in the hours after the storm. Noting that the dead passengers' watches had stopped at 9:30, he backtracked along the currents for twelve hours, re-creating the drift path of beached bodies, empty ice cream canisters, life jackets, deck timbers, and stateroom doors. His nautical sleuthing led shipwreck explorers to the steamer's final resting place off Gloucester, Massachusetts. The *Portland*'s remains, now listed on the National Register of Historic Places, lie protected in the waters of the Stellwagen Bank National Marine Sanctuary.

Sea ice in the Franklin Strait (above) and in other narrow channels in arctic Canada stymied nineteenth-century European explorers seeking a way through the Northwest Passage.

See Map 10, Surface Currents, p. 270; and Map 11, Ocean Conveyor Belt, p. 270.

Hatchling green sea turtles ride ocean currents from the sands of their birth to feed in floating beds of Sargassum weed.

Along the Surface Drift Passengers Great and Small

The journey of the currents is unending. Water descending into the depths returns and flows along the sea surface, where it leaves a visible trail. A seed from the Amazon rainforest—light, smooth, and hard as a rock—flows with the river to the sea, up through the Caribbean, making landfall on a sea island off the coast of Georgia. A glass float from a Japanese fishing net rides a current across the Pacific to an Oregon beach. A bottle tossed into the stormy Gulf of Maine is carried across the Atlantic to an island in the English Channel. A tropical fish from the Indian Ocean is spun around the Cape of Good Hope and delivered to the Atlantic, where it makes a new home on a Caribbean reef. Seed, float, bottle, fish: these tiny drifters leave one shore or ocean for another, carried by currents for hundreds, sometimes thousands, of miles.

Borne by currents, dismasted ships have drifted across entire oceans. In the fall of 1832 a Japanese junk carrying a tribute of rice and ceramics for the emperor was caught in a typhoon. Heavy winds and high seas tore the sails and broke the rudder and mast, setting the vessel adrift in the Kuroshio Current. It floated helplessly for many months as, one by one, the crew died of hunger and exposure. More than a year later, the ship and its three survivors washed ashore on Washington's Olympic Peninsula. In all likelihood, it was not the only Japanese vessel to ride the current across the Pacific.

In the ruins of an ancient Native American whaling village on the tip of the Olympic Peninsula, anthropologists have found pieces of bamboo and iron blades used in chisels and adzes. They speculate that currents carrying Japanese shipwrecks across the Pacific may have brought iron to the Ozette people living at the site, long before Europeans arrived. Other sailors crossed wide swaths of the Pacific intentionally.

Restless motions of seafloor built the islands of the Pacific, providing evidence that corroborates legends of Polynesian seafarers sailing from Tahiti to Hawaii and back a thousand years ago—a round trip of about 5,000 miles (8,000 km). These highly skilled long-distance mariners hewed their wooden canoes with stone adzes made from the islands' volcanic rock. Scientists analyzing chemical signatures in an adze from the Tuamoto Islands in French Polynesia, east of Tahiti, traced its geologic source to Hawaii, confirming the ancient stories.

Drifters and seafarers, though, are not the only passengers to ride the currents. Equatorial waters, heated by the blazing sun, are pushed by the trade winds up against the edges of continents. There they are deflected poleward by the rotating earth. Narrow and swift, these currents—the Agulhas and Brazil currents in the Southern Hemisphere, the Kuroshio Current and the Gulf Stream in the Northern—carry a massive volume of water: the Gulf Stream, traveling 100 miles (160 km) each day, carries 300 times more water than the Amazon, earth's greatest river. Its most precious cargo is heat.

The Gulf Stream, a ribbon of cobalt blue flowing through a green sea, can be 20 degrees Fahrenheit (11°C) warmer than the surrounding water. Its heat is palpable: winter in London is considerably milder than winter in Labrador, and date palms flourish in Britain's Isles of Scilly when Newfoundland is socked in with snow. Ocean currents heat a harsh earth; the sheer volume of energy they carry has no human-made equivalent. In only five days, the Gulf Stream carries as much energy through the Florida Straits as the entire human world uses in one year.

Warm currents coursing toward the poles deliver heat and then circle back. Broad, cold, slower-moving, they are known by different names—the California Current, the Humboldt Current off Peru, the Benguela Current off Namibia and South Africa. They bear other passengers. As they veer away from shore, deep water rises in their wake, rich in nutrients from organisms that lived and died at the surface, then sank and decayed. Carried from the depths by the pulse of currents, these nutrients create some of the sea's most fertile waters, sustaining rich anchoveta populations off Peru, sardines off California, and the tens of thousands of seabirds and seals that feed upon the fish.

Fertile currents provide abundant food: blue-footed boobies feed on herring off the Galápagos Islands (left); copper sharks attack a sardine bait ball off South Africa (right).

Currents carry heat from the equator and bring nutrients from the depths. These are but two ways the sea provides. There is another. Almost all of earth's water—96 percent—rests in the sea. Three percent is frozen into glaciers and polar ice, and a mere 1 percent fills the lakes, rivers, and groundwater that moisten thirsty continents. The sea, though, is the ultimate source of earth's life-giving water. Most of the rain that falls, falls on the sea. Most of the water evaporated to make rain comes from the sea. Some rises on the wind and is blown onto the land, raining out to fill lakes and rivers and soaking into soil, but its terrestrial stay is temporary. Breathed out by trees or evaporated, it returns to the sky and rains out again, cycling between earth and sky until, at last, it returns to the sea. The journey is long.

A Drop of Water

A drop of water from the Atlantic, evaporated in the heat of the tropics, may blow with the trade winds across Panama and rain into the Pacific, where it might ride the currents to Japan, then loop back across the sea. Along the way, it might be caught in a swirling gyre of garbage washed off land a thousand miles (1,600 km) away. It might circle with the plastic for dozens of years, then spin back into the currents to ride west through Indonesia into the Indian Ocean.

There, evaporated from warm water, it may be carried by wind, falling in monsoon rains to water a rice field in Bangladesh and flow with the Ganges back into the sea. Somewhere off the coast of Madagascar, it may ride the Agulhas Current into a rogue wave off the coast of South Africa, then take an icy detour around Antarctica in the ocean's longest and most powerful current. At the edge of the sea ice, it may be sipped by tiny krill or filtered through the baleen of feeding humpback whales. In time, the drop may exit the Antarctic Current and head north. Off the coast of Peru, it may be swallowed by a schooling anchoveta that in turn is swallowed by a cormorant that excretes it as guano on Peru's rock islands. In the hot sun, the drop may evaporate and return to the sea as rain, floating north to circle the Pacific once again.

Nutrients in the ocean's currents provide food for tiny krill (top, off Antarctica) and humpback whales (bottom, bubble-net feeding off Alaska).

This time, on a turn south with the California Current, it might evaporate and rain out into the Olympic forests of the Pacific Northwest. Sucked up by the roots of a Sitka spruce or Douglas fir and pumped up to the crown of the tree, it might once again take to the air. Most likely, it would rain out again and wash into the sea, but it just might move east with the wind, crossing mountains and plains, perhaps passing through a big cottonwood or a delicate prairie flower, and then perhaps through a family's well and into a cool glass of water on a kitchen table before joining a stream that joins the Mississippi and flows back into the ocean. It may take that drop several thousand years to make a full circuit.

Water evaporated from the sea rains out in forests (left, Olympic National Park, Washington), eventually filling rivers (right, Lake Itasca, headwaters of the Mississippi River in Minnesota),

Villagers planting a rice paddy along the Ganges, whales feeding near Arctic sea ice, giant turbines generating electricity at the Hoover Dam, thirsty urban dwellers turning on their faucets—all depend on water that came from the ocean. Seawater flows in currents with no beginning and no end, indivisible, sustaining squid and schoolboy, maple and mountain lion, cormorant and cod. Flowing water is timeless: the journey of currents, from tropics to icy polar seas and back again, continues as it has for four billion years, ever since the first water evaporated from the first scalding sea and fell as rain.

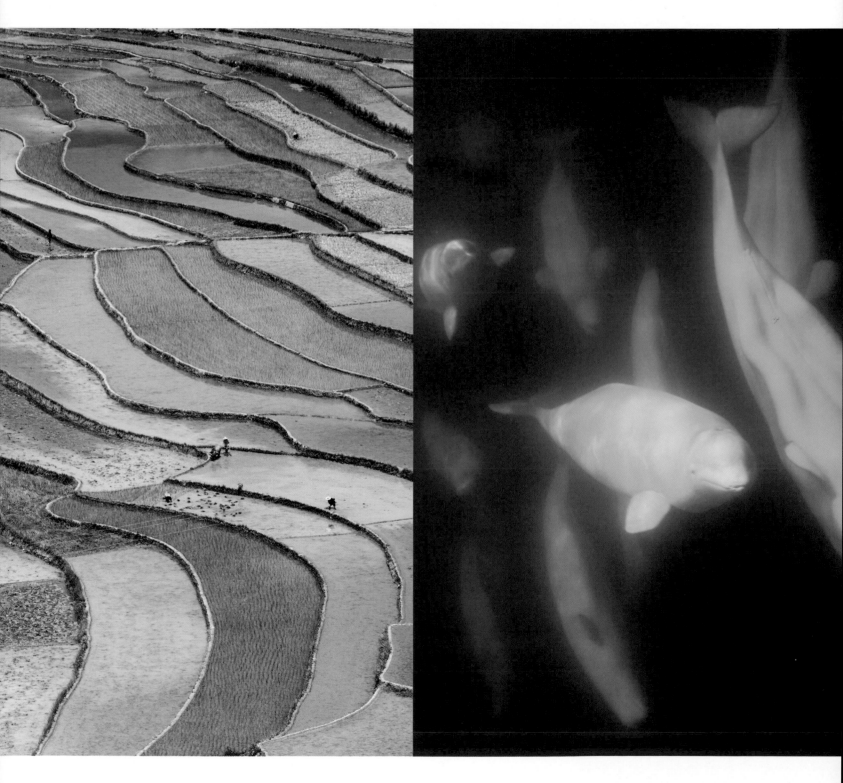

fertilizing fields (left, rice paddies along the Chang Jiang, China), and returning to the sea (right, beluga whales at the mouth of the Churchill River in Canada's Hudson Bay).

Connections:
Opening and Closing Seas, Flowing Currents

Yellowfin goatfish (Mulloidichthys vanicolensis) school along Australia's Great Barrier reef. Their larvae have traveled across the Pacific from one shallow-water habitat to another, riding the currents across one of the world's widest deepwater barriers.

The Pacific is a closing sea, raising the high Andes and Rocky Mountains as its basin crumples beneath the American continent. These mountains, aided by the press of wind and the flow of water, sculpt both rainforest and desert. Wind laden with moisture from the Pacific blows into the towering Cascades and yields its water, each year drenching the lush, wet Olympic Rainforest with 200 inches (500 cm) of rain. Farther south, copious amounts of seawater, lifted on the trade winds to the Amazonian rainforest, are blocked by the Andes. Years pass with no precipitation in Chile's Atacama Desert.

A closing ocean basin redirects the rain and strengthens currents, with startling and profound effects. Five million years ago, as the Caribbean narrowed, North and South America pivoted toward each other, lifting the Isthmus of Panama. The narrow isthmus closed the gateway to the Pacific, turned the Gulf Stream north, and divided the sea. Snapping shrimp that once mingled and mated in the tropical seaway joining the Atlantic and the Pacific separated and evolved into bellicose strangers. Currents altered a Caribbean neighborhood that was once more homogenous: upwelling currents cooled and fertilized the Pacific, attracting coldwater fish, while staghorn and elkhorn coral grew in the warm, calmer Atlantic.

Elsewhere, currents connect a sea that drifting continents divide. When Africa collided with India, the ancient Tethys Sea closed, leaving a massive land barrier between the Atlantic and Indian oceans. Rushing currents keep these waters joined, carrying tiny reef fish thousands of miles from one ocean basin to another. Pale goldspot gobies and bright orange pygmy angelfish living in coral reefs and rubble off the coast of Durban, South Africa, floated into the Atlantic on eddies spun from the Indian Ocean's warm Agulhas Current. The Atlantic's cold Benguela Current is lethal to tropical fish, but gobies and angelfish, sheltered in the warm eddies, drifted north and then crossed the sea to settle in Brazil and the Caribbean.

This transoceanic journey, begun 250,000 years ago for pygmy angelfish, 130,000 years ago for goldspot gobies, continues, aided by mountains rising from the sea as the Atlantic widens. Two peaks from the Mid-Atlantic Ridge, St. Helena and Ascension, rising in solitude above the waves, create shallow-water stepping-stones for reef fish following the currents. In the last twenty-five years, scientists have found large numbers of fish in warming water off the Canary Islands.

Elkhorn coral with sturdy, antlerlike branches that grow 2 to 4 inches (5–10 cm) per year once dominated shallow-water coral reefs in southern Florida and the Caribbean. It is now listed as threatened under the U.S. Endangered Species Act.

Deluge

In India and Bangladesh, millions of lives depend on seasonal rain brought from the sea. During the sunny, clear winters, dry, cool winds blow down from the mountains out to sea; come summer, they reverse. Hot, dusty air rises off the baking land. Humid air, heavy with moisture from the Indian Ocean, replaces it. The eagerly awaited monsoons flood ponds and fields, wells and reservoirs. Swollen rivers carry silt down from the mountains, restoring fertility to well-worked soil, turning parched land green. The rain brings water for drinking, for irrigating crops, for survival. It is not gentle.

Monsoon rains fall hard and fast—almost an entire year's worth in a few months— flooding entire villages, turning roads into rivers, taxi drivers into boatmen, and sending people to their rooftops. On July 26, 2005, 26 inches (66 cm) of rain inundated Mumbai (formerly Bombay), the most rain ever recorded in India within a single day.

Tropical seas moisten the wind, warming it, delivering rain to dry lands, blasting heat away from the scorching tropics. The work can be violent. Throughout the world, hurricanes are named for the wind: Huracan, Caribbean god of the wind; *taifeng*, a great wind in China; Typhon, Greek father of the winds; Hurakan, Heart of Heaven, a Maya god present at the creation, who from a misty cloud called the earth from the sea.

Though named for the wind, hurricanes are spawned only from the sea. They arise in late summer, when the sea is at its warmest, and winds begin to spiral around areas of low pressure. Water evaporates, rises with the wind, and condenses. The rain releases heat, powering more wind that captures more water, then rises, rains, and releases more energy. The hurricane or typhoon feeds itself, wreaking havoc as it makes landfall.

At the edge of a warm sea—along the Gulf Coast of the United States, in Bangladesh and China, and in Honduras and Nicaragua—wind, rain, and tidal surges from typhoons and hurricanes have killed thousands of people and rendered thousands more homeless, destroying villages and cities and ruining rice fields that are the primary source of food for many people. No other storms are so destructive.

Hurricanes begin at the sea surface, but they can reach down to the seafloor, scouring a coral reef or churning the bottom into a speeding current of mud, silt, and sand. When the debris settles, it buries whatever bottom communities lie in its path. Death is instantaneous, but a record of the life can endure. A little over 500 million years ago, a hurricane swept through a shallow sea, smothering thousands of animals in swirl-ing mud. Time passed, and the muddy tomb hardened into stony seafloor. More time passed, and the water drained away, exposing the hardened rock. When it cracked, it revealed the record of life in an ancient sea, a sea that gave rise to earth's first animals.

Explosion of Life

Life appeared on earth approximately three and a half billion years ago. Some two and a half billion years ago, bacteria floating in earth's ocean began breathing oxygen into the atmosphere. Though they were invisible to the naked eye and their lives were short—a matter of hours or, at most, days—their legacy endures. Breath by breath, they altered the atmosphere, opening the way for a burst of life that had never been seen before, and has never been seen since. Relative to the evolutionary quiet that preceded it, the first large animals (if indeed they were animals) emerged swiftly, seemingly from nowhere.

An Enigma

Newfoundland's Avalon Peninsula, at the southeastern tip of the island, juts out into the sea, a rocky, desolate, windswept point battered by storms and waves, and often shrouded in fog. Out on the point is a reconstruction of the old Marconi wireless station that, on the night of December 14, 1912, received news of the *Titanic*'s fatal collision with an iceberg. Basque, French, and English fishermen, sailing across the Atlantic for cod in the 1500s, built a seasonal fishing station at nearby Portugal Cove South. Later, when the village was inhabited year-round, its residents earned their living from the sea, catching cod, then salting and drying it on racks on the beach. In 1991 the Grand Banks cod fishery collapsed, threatening a way of life with extinction. Though the fish are gone and the fishermen are seeking other livelihoods, the headland remains, testament to an earlier time in earth's history, when another way of life flourished and disappeared.

Background: Pressed into the seaside cliffs of Mistaken Point, Newfoundland, are fossils of earth's first large organisms, the Ediacara, which once lived on the bottom of an ancient ocean.

Wind, rain, and salt spray have scoured the headland, eroding the softer rock, exposing thousands of fossils. They are large, some over 6 feet (2 m) long. Shaped like fern fronds, spindles, and disks, they are odd and unfamiliar. They are the oldest large, complex fossils known anywhere in the world, but their identities continue to elude scientists. What they were, how they lived, and where they came from are mysteries still to be solved; they share little affinity with inhabitants of today's ocean and don't easily fit into any of today's known phyla.

Ediacaran fossils are found in ancient seafloor throughout the world, in Newfoundland, Australia, Namibia, China, and the White Sea area of Russia. Above left: Tribrachidium, Ediacara Hills, Australia. Center left: Rangeomorph, Spaniards Bay, Newfoundland. Center right: Spriggina, Ediacara Hills, Australia. Right: Dickinsonia, Ediacara Hills, Australia.

Flinders Ranges, Australia: Many Ediacaran fossils are preserved in the Ediacara Hills of Australia's Flinders Ranges.

Now cast in stone and exposed to the sun, some 570 million years ago they lived in deep water below the crashing waves, beyond the reach of light. Tethered to the ocean floor by holdfasts, like seaweed clinging to rock, their bodies swayed with passing currents. Bottom currents were their lifeline, bringing oxygen and sustenance into barren water. Whether they simply absorbed nourishment, filtered it, or were assisted by symbiotic bacteria, no one really knows. They had no teeth, no mouths, no guts.

One day whole clusters of them died, all at once. They were lying on the seafloor, almost fully prone, bending before a current. A volcano onshore exploded, the rain of ash settled in the sea, and they were buried. None of the organisms were ripped apart, none tumbled down the slope; it was a gentle death. They died where they lived, preserved in exquisite detail.

The animals had soft bodies, and often when they were buried, they collapsed. A little farther up the Avalon Peninsula, along the rocky coast of Spaniard's Bay, where they have been preserved in three dimensions, their bodies resemble quilted air mattresses. The delicate design, which defies easy interpretation, is unknown in any modern organism.

Ediacaran fossils, as they are called, are found all over the world: in the desert in Namibia; the English midlands; the Ediacara Hills in Australia, for which they are named; the White Sea in Russia; and the Yangtze Gorges. Their stay, thirty million years, was short relative to earth's early and enduring photosynthesizing bacteria. The architecture of the leaflike, quilted *Charnia*, the mysterious *Tribrachidium*, with its highly unusual three-part symmetry, and the ellipsoidal and controversial *Dickinsonia* were evolutionary roads not taken, way stations on a path never completed. *Kimberella* and *Spriggina* also defy easy classification, and although they too are long extinct, their legacy of bilateral symmetry endures in our own bodies. Soft and stationary, without means to protect or defend themselves or to flee, the gentle Ediacara were unprepared for the world to come. Long before that world appeared, there were glimmers, inklings of potential for momentous change.

Glimmers of a World to Come

In the Top End region of northern Australia and on Victoria Island in western Canada, remnants of old seafloor hold fossils of *Tappania*, a small organism of mysterious affinities. One and a half billion years old, its branching cells vary in size, each with asymmetric, irregularly arrayed spikes or spines. Its identity is vague: whether it was an ancestral plant, fungus, or animal is unresolved. Whatever it was, its construction, requiring the internal, flexible scaffolding of a cytoskeleton, enabled it to easily change its own shape. The application of this design, fundamental to multicellular life, would be surprising and unexpected.

Today the Pacific Ocean is closing. Its basin drifts under North America, raising the Rocky Mountains. The Colorado River cuts through the uplifted rock, creating the Grand Canyon, and exposing two billion years of North America's geologic history, including a time 700 million years ago when much of North America was covered by a shallow sea. Deep in the canyon is a layer of that ancient seafloor, packed with thousands of tiny vase-shaped fossils new to the fossil record.

Unlike the sea's plants, which manufactured their own food from sunlight and water, these microscopic organisms were consumers. Instead of making their own food, they ate what others produced, building the first layers of a marine food web. They ate, but were eaten as well. The fossils show evidence of predation, signaling the emerging contest between hunter and prey. In their honeycomb-textured cell walls, they grew tiny mineralized scales, harbingers of a time when sea animals would shelter their soft bodies in a protective hard shell. About 100 species of these organisms, testate amoebae, live in the sea today, almost identical to their ancestors.

Fossils embedded in the walls of a phosphate quarry in southern China and in the gorges of the Chang Jiang (Yangtze River) yield additional evidence that the dawn of animal life was soon to break. That evidence, too, comes from an ancient sea, where

today's rules for earthly engagement were beginning to set. The Doushantuo fossils are highly ornamented; spikes or spines protrude from their walls. Within are embryos in the process of dividing: single cells, and cells that have divided once, twice, four times. They look like miniature soccer balls. No one yet knows what animal they would have become, whether ultimately the dividing cells would have specialized and differentiated into the jointed body of an arthropod, the rounded body of a jellyfish, or the frond of a feathery Ediacaran. They died before they were born, but in their death they captured the birth of the animal kingdom.

A rocky mesa rises above the dust of Namibia. It is a limestone reef built 550 million years ago, when this part of Africa was covered by a shallow sea. The water has long since drained away, and the reef builders themselves are extinct. They bear no resemblance to modern animals. The remnants of *Cloudina* look like funnels nesting one inside another, and *Namacalathus* resembles wine goblets. They left no descendants, only an abundance of shells.

Above: Tiger Leaping Gorge, Chang Jiang (Yangtze River). Fossils embedded in the gorges of the Chang Jiang mark the dawn of animal life.

Right: These fossil embryos from southern China represent some of earth's earliest animals.

Deep in the Grand Canyon of northwest Arizona are tiny vase-shaped fossils that record the first contests between predator and prey.

Desolate reefs in Namibia and steep, inaccessible cliffs on Salient Mountain, high in the Canadian Rockies, are packed with them. Though light, thin, and flexible, and unlike the kinds of shells you'd find on a beach today, they were shells nonetheless. The bodies inside have long disintegrated, but the house is of greater interest than its occupant. The arrival of shells profoundly altered the course of evolution and the ecology of the sea.

By the time the Namibian reefs were built, most of the biological pieces—the developmental "toolkit"—were in place to build a large animal. The animals themselves, though, were a long time in coming. More than a billion years elapsed between the time *Tappania* dwelled in ancient estuaries and deltas and the advent of large marine animals. The possibility appeared early on, but only when the sea itself changed could that possibility be realized, could water that had spawned life's earliest iteration also give rise to large marine animals.

Air and Nourishment from the Sea

Beneath the animal embryos in China, the shelly reefs in Namibia, and the Ediacaran fossils in Newfoundland lie layers of jumbled rock, boulders, cobbles, and pebbles of all sizes mixed together, dumped from a melting glacier that froze every continent and chilled the sea. As the ice thawed, the Ediacara appeared. Analyses of ocean floor in Newfoundland's Avalon Peninsula and in the oil fields of Oman reveal that when the glaciers retreated, the deep sea had finally filled with enough oxygen to support the high-energy metabolism of large animals. That sea is the sea of today.

With sufficient oxygen, an animal can become large. The sea filled not only with oxygen but also with nutrients contributed by closing ocean basins and colliding continents. Approximately 650 million years ago, Antarctica, Australia, and India merged with Africa and South America to raise a chain of high mountains 5,000 miles (8,000 km) long by 600 miles (1,000 km) wide. The mountains eroded, washing enough sediment into the sea to cover the mainland of the United States with a layer of sand and silt about 6 miles (10 km) thick. The sediment was nutritious, containing the metal molybdenum, which would allow growing organisms to process nitrogen. It also contained iron and phosphorous, essential for the growth of voluminous amounts of algae animals would eat, and calcium, which would saturate the sea, enabling them to manufacture shells.

Earth and the life it nourishes evolve together, each in need of the other, each shaping and reshaping the other. When the sea had filled with air to breathe and sufficient nourishment, animal life burst forth.

Above, blue shark: Most animal body plans appeared in the oxygenated and nourishment-filled Cambrian ocean, including animals with backbones. From them, millions of years later, sharks would evolve.

Fossils in the Burgess Shale high in the Canadian Rockies record the rise of earth's first large animals: Marrella, Pikaia, Hallucigenia, Anomalocaris, Ottoia, Opabinia (below, left to right), and Olenoides (opposite, above).

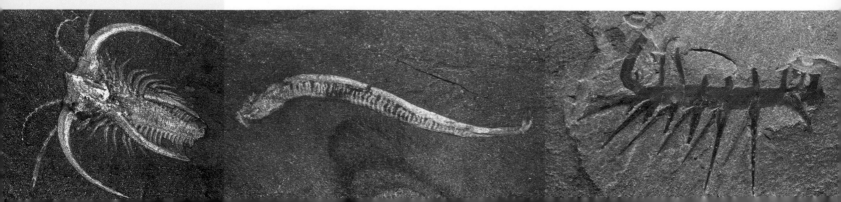

The Blossoming of Animal Life

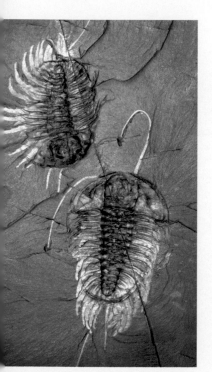

When marine animals finally arrived, they arrived in force. Some were fierce, huge, and bizarre, their place on the tree of life still dimly understood. Fossils of *Anomalocaris*, the "odd shrimp," looked somewhat like a shrimp but then turned out to be nothing of the kind. Up to 6.5 feet (2 m) long, this underwater giant was a top predator that grasped its food with two long, spiked arms and shoveled it down a gaping orifice. The mouth, which resembled a pineapple ring, held two or three circles of jagged teeth. Two side flaps enabled the monster to glide through the water as it searched for food.

Scientists had difficulty piecing together so large and unfamiliar a creature. Initially, they believed its many parts belonged to separate, smaller animals: the mouth to a primordial jellyfish, the body to a flattened sea cucumber or sponge. The complete animal, when it was assembled, was frightening and, given the passive Ediacara that came before, unanticipated. In *Anomalocaris*, multicellularity had burst into full expression: the animal had muscular, nervous, digestive, and circulatory systems. It had fully differentiated and specialized body parts: legs, flaps that were somewhat like fins, a head, feeding appendages, mouth, teeth, a gut, brains.

A confluence of circumstances prepared the way for the startling and spectacular arrival of animals in the Cambrian period (as this time was called). Each was essential; none alone was sufficient. The genetic toolkit, enabling embryonic cells to develop into specialized limbs and muscles, made the evolution of animals possible. The infusion of the sea with oxygen and minerals created a nurturing environment where that evolution could occur. The animals themselves precipitated the sudden and extraordinary diversity of life, profoundly reorganizing the structure of life in the sea, giving birth to the modern ocean.

Cambrian fossils open a window into this unparalleled time of evolutionary innovation. The fossils come from animals that dwelled in warm, shallow water, and that were buried alive, smothered by mud and silt stirred up by hurricanes or benthic storms. The mud turned to rock, the water drained away, and the seafloor was pushed up into continents. The continents have drifted far from their Cambrian locations near the equator, but the fossils remained intact, beautifully preserved in their mud graves. Today they lie amid the hills and lakes in and near Chengjiang, China; in Sirius Passet, in remote northern Greenland, at the edge of the Arctic Ocean; and in the Burgess Shale, high in the Canadian Rockies. Thousands of fossils representing hundreds of species lie in the exposed rock, charting the rise of earth's first large animals.

Lands Divided and Joined

Just as closing ocean basins and continents helped create an environment that permitted the evolution of animals, so their fossils, and the rocks that house them, help reconstruct that ancient ocean and continents, mapping a lost world. Today, on the western side of Newfoundland, the Long Mountains overlook the chilly waters of the Gulf of St. Lawrence, and on the east, the Avalon Peninsula reaches into the icy Atlantic. Five hundred million years ago, they both edged a now vanished tropical ocean near the equator, but they weren't joined as they are today. Fossils from that ancient sea indicate that the two pieces of Newfoundland, separated by only a few hundred miles now, once stood on opposite shores across a wide sea.

Broom Point in Gros Morne National Park (below) and the Avalon Peninsula (opposite) both lie in Newfoundland. Their fossils indicate that they once stood on opposite shores of a wide sea.

The Cambrian ocean teemed with trilobites, arthropods with jointed legs and hard carapaces. The carapaces made excellent fossils. Trilobite fossils in western Newfoundland, the Burgess Shale, and many other locations in the United States and Canada represent animals that lived in the coastal waters of the old continent of Laurentia. In eastern Newfoundland's Avalon Peninsula, in the gorges of the Manuels River near Conception Bay, are fossils of *Paradoxides*, a trilobite that belongs not to Laurentia but to another place altogether. Its contemporaries have been found in abandoned quarries of Braintree, Massachusetts, high in the rugged coastal cliffs of St. David's, Wales, and in the Atlas mountains of Morocco.

FOSSILS MAP DRIFTING CONTINENTS

Though scattered today, the realm of *Paradoxides* was once whole. *Paradoxides* lived in the coastal waters of an ancient arc of islands in the Southern Hemisphere, across the Iapetus Ocean from Laurentia and just off the mighty continent of Gondwana. When Iapetus closed, the islands were crushed between the two landmasses. Millions of years later, the Atlantic opened and the former neighbors were dispersed, some to join Newfoundland and New England, others destined for Europe and Africa. Opening and closing ocean basins fragmented this ancient land, dispersing its pieces, but *Paradoxides* retains the record of its former wholeness.

CONFIGURATION OF CONTINENTS
LATE CAMBRIAN
514 MILLION YEARS AGO

Ancient Landmass

Modern Landmass

Subduction Zone
(triangles point in the direction
of subduction)

Seafloor Spreading Ridge

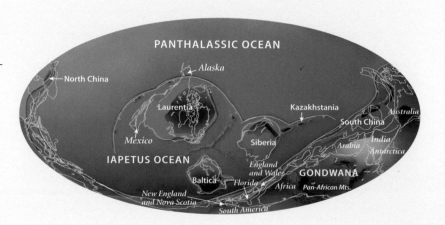

Trilobites developed remarkable adaptations to stalk prey or to avoid being eaten. *Above and right: The sharp spines of* Hoplolichas plautini *(from the Wolchow River basin, St. Petersburg, Russia) made this trilobite unattractive to potential predators.*

A Sea Both Strange and Familiar

Above: With its extended eye stalks, the trilobite Asaphus kowalewskii *(from the Wolchow River basin, St. Petersburg, Russia) burrowed in the mud to evade predators or to await prey.*

Algae and bacteria, using sunlight, oxygen, and chlorophyll to manufacture carbohydrate, provided the critical foundation for life in the Cambrian ocean. Photosynthesizing organisms making their own food had been the sea's primary producers for the previous two billion years. They still are today; almost all animal life depends upon them. The Cambrian ocean swarmed with consumers engaged in a life-or-death struggle for food. As they preyed upon each other and fought to escape predation, they developed tougher shells, the means to crawl and swim, and novel ways of feeding. They evolved together, responding to their own innovations. Competition quickly produced almost every animal body plan known today, and created the complex, layered food webs that, 500 hundred million years later, continue to sustain life in the sea.

Sponges pumped seawater through their porous, vase-shaped bodies to filter out food. Anchored to the bottom, they developed their own ecological niches; they came in many sizes, each feeding at different heights above the seafloor. Many Cambrian sponges became extinct, but the glass sponge, with its thin, needlelike skeleton, lives on as it did millions of years ago. Jellyfish also trace their ancestry to the Cambrian ocean, where they trapped their food with long tentacles as they do today.

Worms evolved, and altered the character of the sea bottom. They burrowed, seeking food and shelter. Once they began burrowing, the hard bacterial mats covering the seafloor were grazed away, and the bottom turned soft and soupy. Worms ate and were eaten; remnants of other animals are found in their intestines, and they themselves were devoured by trilobites, distant relatives of modern scorpions and horseshoe crabs. Trilobites had segmented bodies, many pairs of jointed legs, and antennae. They burrowed in the sediment, crawled on the seafloor, swam. Some were passive feeders. Others hunted.

The diversity of trilobites was extraordinary. Some were blind, others were among the earliest animals to develop complex eyes. Some eyes rested on stalks, enabling the animal to peer above the mud. Compound eyes with many lenses helped swimming trilobites find prey and mates in the three-dimensional world of water. When danger lurked, they could roll themselves up in a protective ball like modern pill bugs or deter predators with their sharp spines. They were better protected than Ediacara resting passively on the seafloor. Yet even with all these defenses, they were still vulnerable.

Many dwellers of the Cambrian ocean are now extinct, but their relatives endure. Below: An azure vase sponge (Callyspongia plicifera) *with a brittle star* (Ophiothrix suensonii). *Right: Jellyfish* (Cyanea *sp.*), *Bay of Plenty, New Zealand. Opposite: Horseshoe crabs* (Limulus polyphemus) *spawning, Delaware Bay, New Jersey.*

In newly developing Cambrian food webs, comb jellies ate chaetognaths (arrow worms), snaring them on sticky tentacles. Trilobites had a varied diet; some preyed on worms, some on other trilobites, and some sifted organic particles from the soft mud of the seafloor. *Anomalocaris*, at the top of the food web, may have been the largest and fiercest predator of them all. A few trilobite fossils bear bite marks from a presumed *Anomalocaris* attack, and pieces of fossilized excrement (whose size implicates *Anomalocaris* as their producer) contain trilobite fragments.

Anomalocaris lived at the top of a rich food web that included plants, bacteria, herbivores, carnivores, and omnivores. Marine food webs endure today, transferring food and energy from the tiniest plant to the largest whale, although many of the Cambrian constituents have long disappeared. *Anomalocaris*, an apex predator, is extinct, but descendants of other, smaller, less conspicuous Cambrian animals rose to take its place. One of them was an ancestral fish.

A Green Sea

A school of blue maomao (Scorpis violacea) swims above a bed of kelp, Poor Knights Islands, New Zealand.

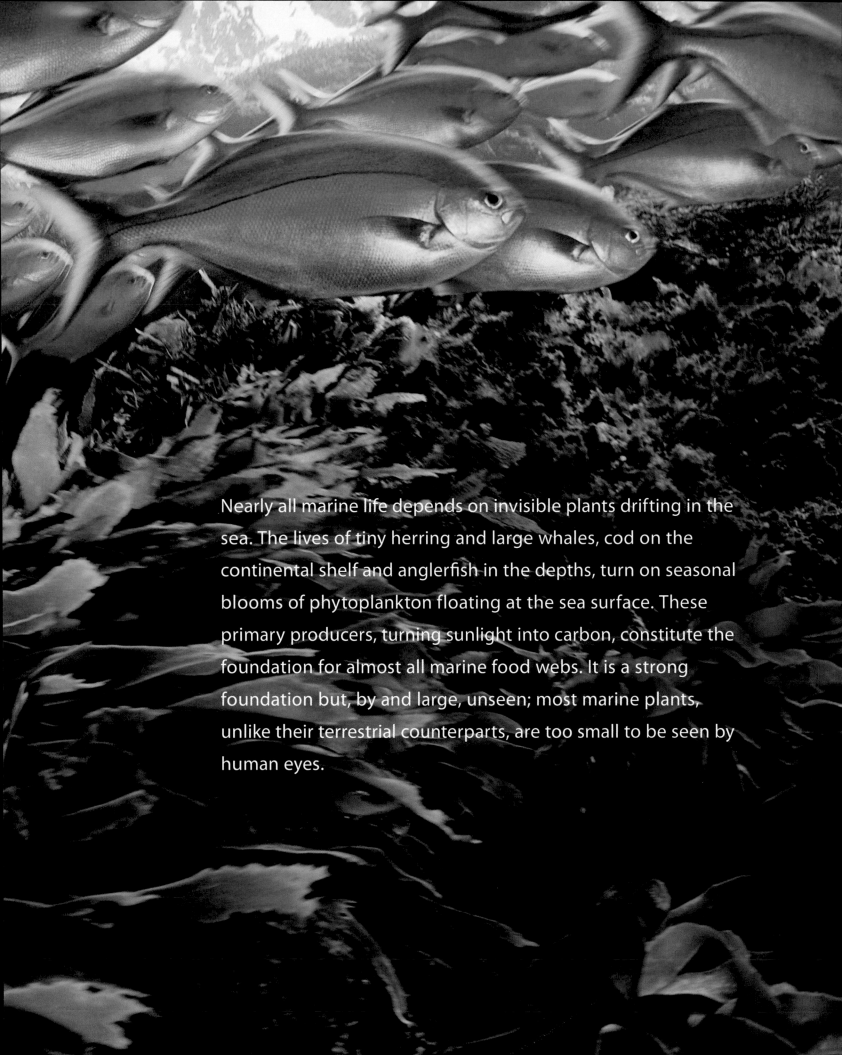

Nearly all marine life depends on invisible plants drifting in the sea. The lives of tiny herring and large whales, cod on the continental shelf and anglerfish in the depths, turn on seasonal blooms of phytoplankton floating at the sea surface. These primary producers, turning sunlight into carbon, constitute the foundation for almost all marine food webs. It is a strong foundation but, by and large, unseen; most marine plants, unlike their terrestrial counterparts, are too small to be seen by human eyes.

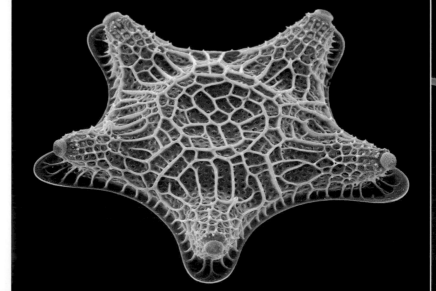

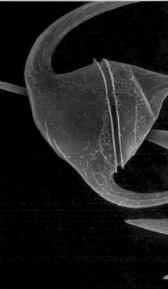

Phytoplankton, single-celled organisms living in the sea and capturing energy from sunlight, produce almost half earth's oxygen. When magnified by scanning electron microscopes, their immense variety is revealed. Above: Pennate (Diatoma) *and centric* (Coscinodiscus) *marine diatoms. Right: marine diatom* (Triceratium). *Far right: Marine dinofla-gellates* (Ceratium tripos, C. furca, *and* C. fusus).

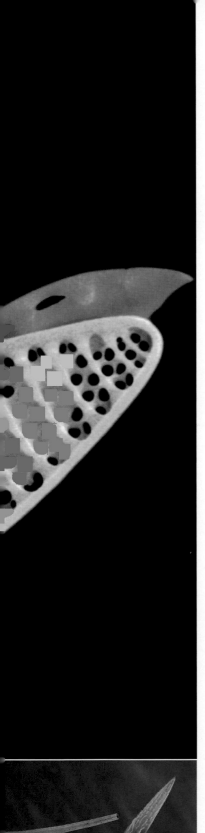

The Power of Phytoplankton

Single cells of chlorophyll breathed oxygen into ocean and atmosphere two and a half billion years ago. Their descendants still endure. Other floating plants of the modern ocean began evolving more recently, around 200 million years ago, as the emerging Atlantic flooded the land and widened the continental shelves, and as life began to recover from a devastating mass extinction. Satellites, scanning the extent of photosynthesis across entire oceans, reveal the monumental importance of the sea's drifting plants: constituting a mere 0.2 percent of earth's plant biomass, they carry out almost half the planet's photosynthesis, keeping the atmosphere supplied with oxygen.

Their individual lives are fleeting. The entire mass of marine phytoplankton replaces itself every few days, compared to an average turnover time for terrestrial plants of nineteen years. These drifting plants are small in size but great in number. Billions upon billions of coccolithophores may bloom at once, turning a swath of sea the size of England a milky white. Individual plants are covered with plates that resemble clusters of translucent pineapple rings. Made of chalk, these elegant plants may make more calcium carbonate than any other organism. When they die, some drift to the seafloor, where they pile up ever so slowly, layer after layer, year after year. When the sea recedes, what once belonged to the ocean becomes land. The white cliffs of Dover, England, and France's champagne cellars are carved from the remains of coccolithophores that piled up on the seafloor and turned to rock.

Among the ocean's most abundant producers of carbon are diatoms, single-celled plants with glassy, silica skeletons. At the end of winter, when storms have thoroughly mixed the sea, bringing up rich nutrients from the depths, and the days begin to lengthen, diatoms bloom, turning winter's gray sea green. They live singly or in chains and divide rapidly, a single cell producing a million progeny within three weeks. As summer continues, the water grows warm and calm. Diatoms use up the sea's stores of silica, and their populations fade. Then other marine plants, the dinoflagellates—small, single cells ornamented with spines and wings—bloom. Sometimes they luminesce. On dark nights, their faint light may flash in breaking waves or on the wet sand. They produce carbon throughout the summer, sustaining a sea of plenty.

The Biology of the Very Small

The plenty quickly disappears. Half immediately goes to feed larger plant-eating animals, and half is routed through food webs of even tinier organisms whose existence we are only beginning to realize. Marine bacteria soak up carbon "leaked" from the sea's drifting plants, creating additional meals for marine organisms that do not make their own carbohydrate. Bacteria decompose waste from other organisms as it falls to the depths as "marine snow," releasing nutrients for reuse by floating plants. They even swim in to attack living plants. They are the sea's great recyclers, wasting nothing, releasing minerals from the dead to nourish the living, again and again.

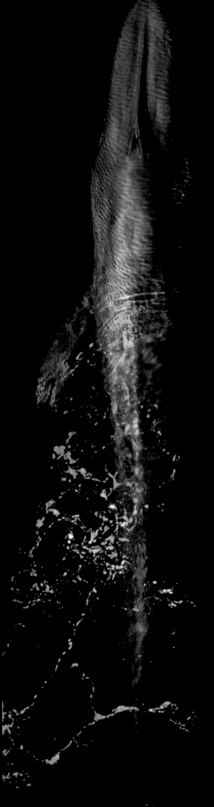

Their populations do not rise unchecked. They have their own predators—copepods and protozoa, among others, and tiny viruses. Single or double strands of DNA or RNA covered by a protein or lipid coat, viruses are unable to exist on their own. Instead they reproduce inside their host until it bursts, leveling marine bacterial populations by as much as 40 percent every day and transferring their genes across entire oceans.

We hardly know this minute world. On the tree of life, plants and animals that can be seen with the naked eye are mere twigs, a tiny portion of all life on earth. Whether measured by numbers of individuals or volume, most life on our planet exists unseen by us. As scientists develop new ways of "seeing" the ocean, more of this invisible world is revealed. Bathing a stream of water in the light of a laser, scientists found *Prochlorococcus*, the sea's most abundant photosynthesizer. Analyzing archival satellite data, other scientists found corroboration for what mariners from the Indian Ocean have long reported: on moonless nights in calm water, a faint white glow extending to the horizon, making it seem as if their ship were sailing over clouds. Satellite data confirmed the widespread bioluminescent glow, documenting its extent and three-night duration. Scientists believe that billions upon billions of glowing bacteria turned the sea milky white.

Viruses are the most abundant biological entities in the sea, but by and large their identities are unknown. Genomic sequencing of viruses in ocean water has found several hundred thousand, 91 percent of which were new to science. Other sequencing of marine microbes suggests that the sea may harbor between five and ten million species of microorganisms—far more than originally thought. Even these numbers are subject to revision. Each discovery yields new insight, revealing how little we know the sea.

The dwellers of this invisible world come in increasingly smaller sizes and greater numbers than we had imagined. Diatoms and coccolithophores are roughly half the width of a human hair, marine bacteria 100 times smaller, and marine viruses 1,000 times smaller. Making up in numbers what they lack in size, a few drops of water may contain one million bacteria and 100 million viruses. Lined up end to end, the tiny viruses of the sea are so numerous they could cross and recross the Milky Way approximately 100 times.

Viruses are the smallest forms of life in the sea, and blue whales the largest. Yet viruses, tiny as they are, contain more carbon. Between 4,500 and 12,000 blue whales roam the sea today; marine viruses contain as much carbon as 75 million blue whales. Carbon and other nutrients—phosphorous, nitrogen, sulfur, iron—are cycled and recycled up through marine food webs to sustain the sea's animals large and small; only a few layers separate viruses from whales.

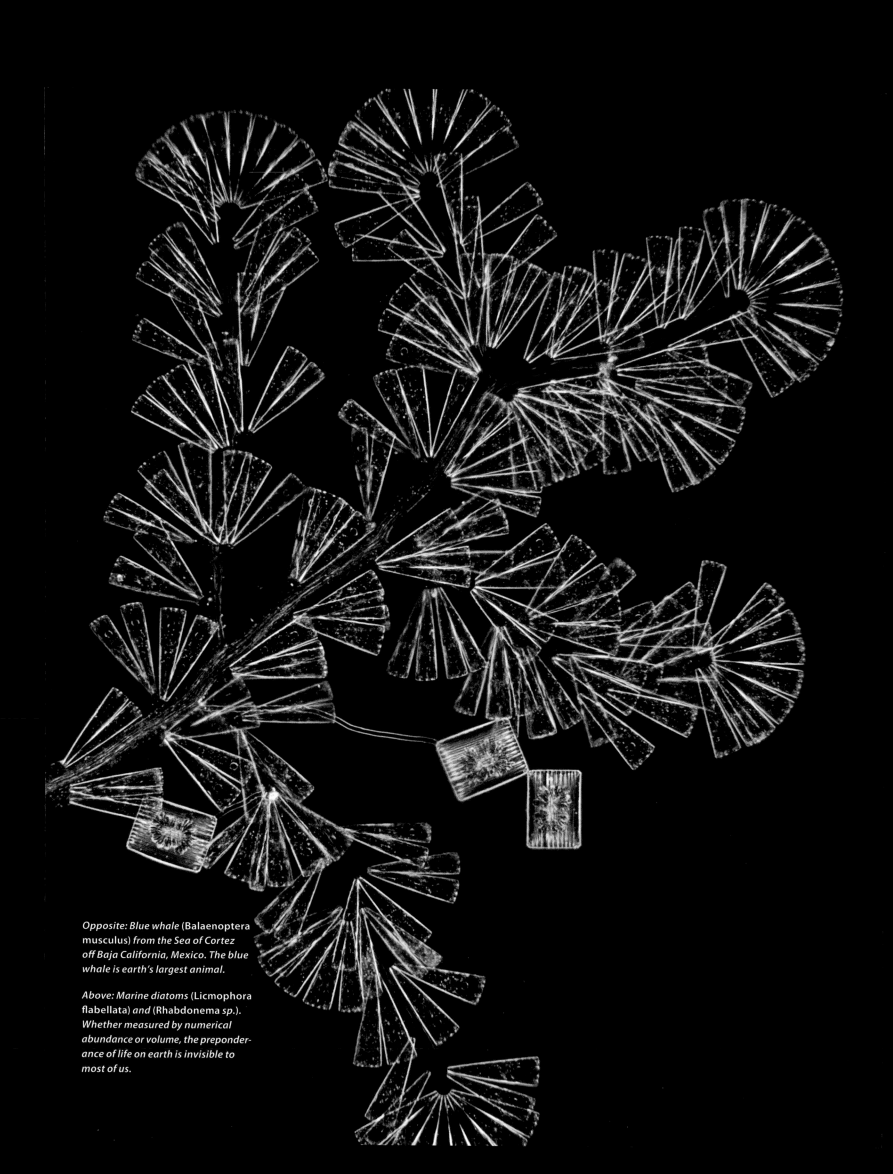

Opposite: Blue whale (Balaenoptera musculus) from the Sea of Cortez off Baja California, Mexico. The blue whale is earth's largest animal.

Above: Marine diatoms (Licmophora flabellata) and (Rhabdonema sp.). Whether measured by numerical abundance or volume, the preponderance of life on earth is invisible to most of us.

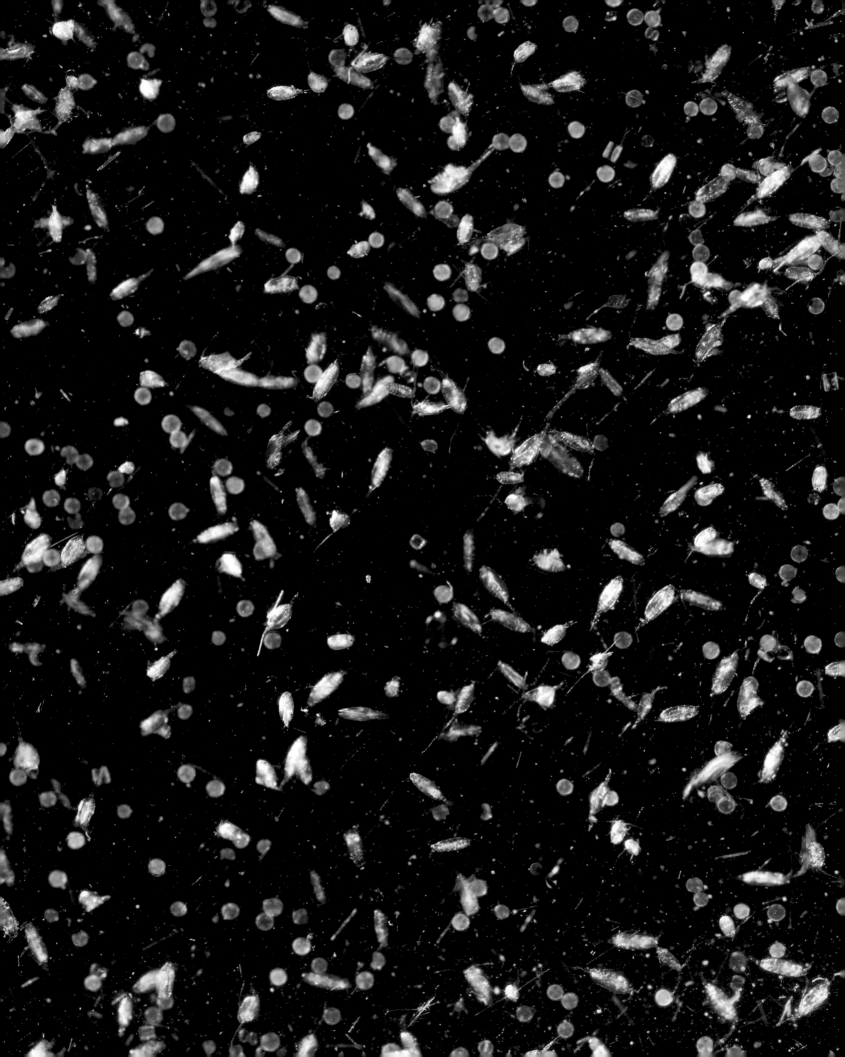

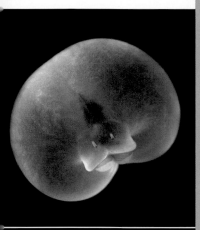

Grazers

Where marine plants bloom, animals quickly swim in to graze. Many, too, are tiny. *Calanus*, a copepod no bigger than a grain of rice, is among the smallest but most abundant animals in the sea. Copepods and shrimplike krill, critical strands in marine food webs, turn the ocean's drifting plants into protein that feeds larger fish and whales. They crop the sea's floating meadows, leaving little behind. After copepods feed, less than 2 percent of the sea's greenery may remain.

The lives of the meadow and its grazers are intertwined. As spring comes to the North Atlantic, diatoms multiply, unimpeded by copepods, whose life cycles are longer and who reproduce more slowly. By the time the first brood of copepods is ready to feed, weeks later, the meadow is in full bloom. In an inlet of the Pacific off Vancouver Island, krill help fertilize the garden they graze, each night swimming up to the sea surface to feed. As the animals make their daily commute, they stir the water. Creating as much turbulence as an incoming tide, they bring up nutrients for the sea's floating plants.

Marine food webs evolved at least 540 million years ago, when the first animals appeared in the sea, and their multiple layers continue to transfer food and energy produced by plants to larger animals. Voracious and streamlined arrow worms, or chaetognaths (named after the Greek *gnathos*, or jaw), were some of the ocean's earliest predators. Their jaws, crowns of grasping spines and teeth, have served them well for millions of years. While other animals have risen to prominence and then disappeared, their places in marine food webs taken by new arrivals, chaetognaths, prodigious consumers of copepods, have survived. Fossil chaetognaths from Chengjiang, China, are virtually identical with their contemporary descendants.

Other grazers may have arrived more recently. A new species of worm, *Chaetopterus pugaporcinus*, living in the waters off California's Monterey Bay, looks like a pig's rump. The worms could be unusually large larvae waiting to metamorphose into elongated adults, or, more likely, they may be formerly sedentary bottom dwellers, newly evolved for a life afloat.

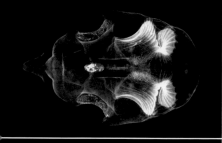

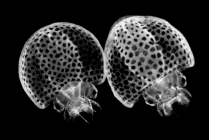

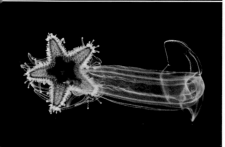

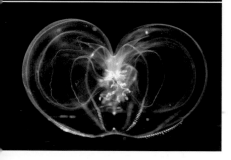

Mostly Water

Buoyed by seawater (800 times denser than air) and bathed in its nutrients, the ocean is home to whole communities of animals that live their lives perpetually afloat. Many that live suspended in the sea are themselves mostly water. When caught in nets towed from a boat, they disintegrate into unrecognizable gelatinous smears. Today remotely operated high-definition video cameras are filming these delicate animals in their native habitats. Some animals live near shore; others dwell in the open ocean. The Census of Marine Life is plumbing the world of zooplankton, and when its work is completed, scientists expect the number of known species to double.

The animals are fragile and beautiful. Translucent sea snails fly through the water with wings, like butterflies. Hovering comb jellies, dark but for their ciliated comb bands, catch prey on sticky tentacles. One, *Thalassocalyce*, the "sea chalice," is shaped like a heart. Colonial siphonophores come in many shapes and sizes: *Praya dubia* may be among the world's longest animals, its stinging tentacles trailing 130 feet (40 m) behind its body, a long, thin string lighting the dark sea. *Athorybia rosacea* is ruffled and round, like a chrysanthemum.

Though their bodies are meager, these delicate animals still provide food to dwellers of the deep sea. *Bathochordaeus* secretes a large, diaphanous house around itself, pumps water through the thin membrane, and filters out food. When its filters clog, it jettisons the entire house and builds another, as often as once a day. The houses, as much as twenty times larger than their inhabitants, are fragile. Falling apart in collecting nets, they had eluded scientists, who now have filmed them, both with their occupants and abandoned, collapsing and sinking to the seafloor. No longer needed by their owners, they become essential nourishment to others.

Advances in technology have opened a window into a world of fragile and beautiful animals living suspended in the water. Above, top to bottom: Larvacean (Oikopleura labradoriensis), squidlings (Thysanoteuthis rhombus), long-armed starfish (Luidia sarsi), and larva of a deep-sea ctenophore (Thalassocalyce sp.). Opposite above: Siphonophore (Nanomia cara).

A SEA WITHIN A SEA

In the middle of the Atlantic lies a sea within a sea. No continent edges its waters; the Sargasso Sea is surrounded entirely by rushing currents. Light falls seemingly undiminished in this lens of warm, clear water 3,300 feet (1,000 m) deep. Compared with fertile waters at the edge of continents, crowded with blooming plants and grazing copepods, the Sargasso is barren but hardly empty.

Eels come to spawn here, swimming hundreds, sometimes thousands, of miles from the icy creeks of Labrador, the bayous of Louisiana, and the upper reaches of the Mississippi. After ten or twenty years of life in freshwater, they find their way back to the Sargasso, to the waters of their birth. The Sargasso holds their secrets; we've yet to follow adult eels on their migration and yet to see them spawn. While eels are visitors to the Sargasso, others reside there year-round.

Golden brown *Sargassum* weed, pushed into long rows by the wind, floats in the water, held up by air bladders the size of berries. Columbus reported rafts of seaweed so wide they reached the horizon, so thick they slowed his ships. The mats, thinner now, still teem with life. Each island of *Sargassum* is a realm unto itself. Barnacles and bryozoans cement their homes to the fronds. Crabs and shrimp find anchorage here, and fish, camouflaged to blend with the foliage. Flying fish spawn here, and chub and snapper larvae grow within the protection of the leaves. Young sea turtles find food and shelter here. Nutrients are few—iron in dust blown off the Sahara, and nitrogen supplied by photosynthesizing bacteria—but they are well used.

Eventually the air bladders thicken, and clumps of seaweed, encrusted with communities they can no longer support, sink. Swimmers depart, and the rest is eaten by dwellers in deeper water. Tips of young *Sargassum*, broken off from older shoots, grow to anchor new floating communities of life.

Opposite: The larvae of eels (Anguilla rostrata) born in the Sargasso Sea drift for at least a year on the Gulf Stream before moving into the rivers of North America and Europe. These young glass eels are entering the Kennebec Estuary, Maine.

Fertile Waters

The sea blooms unevenly. Even where its meadows are thin, overall productivity is high. Most of the sea, 92 percent, is open ocean. These waters, once considered virtual desert, produce three-quarters of the sea's carbon. Where nutrients are more plentiful, the sea blooms profusely. Off the coast of Antarctica, deep water traveling along the seafloor for a thousand years wells up, carrying rich stores of nutrients. As sunlight returns after the long dark winter, and the sea ice begins to melt, diatoms bloom, nourishing dense swarms of krill. Where krill gather, fish, penguins, seals, and whales follow. Adélie penguins forage on krill to feed their young chicks. Giant blue whales feed almost exclusively on krill and copepods, filtering them through their baleen, gulping two-thirds of their body weight with every swallow. The fertility of Antarctica's sea rests on a foundation of invisible floating plants, which each year produce 5.3 ounces (150 g) of carbon for each 10.8 square feet (1 m²) of water.

Adélie penguins (Pygoscelis adeliae), *opposite, and leopard seals* (Hydrurga leptonyx), *below, feed on krill, squid, and fish from the fertile waters off Antarctica. Leopard seals, near the top of Antarctic food webs, feed on penguins as well.*

Copper (Carcharhinus brachyurus) *and blacktip sharks* (Carcharhinus limbatus) *feed in a sardine* (Sardinops sagax) *bait ball in fertile waters off the coast of the Transkei region, South Africa.*

Right: Skeletal remains of marine phytoplankton from the Cretaceous period built the chalk cliffs of the Pinnacles in Studland, Dorset, England.

Far right: Fertile waters support large colonies of seabirds. Here, a gannet colony on Great Barrier Island, Hauraki Gulf, North Island, New Zealand.

Rich as this water is, there is water still more fertile. Coastal waters make up less than 8 percent of the ocean, yet they constitute earth's most productive seas, yielding 90 percent of the catch of the world's fisheries. As trade winds blowing along the coast of Peru push the current offshore, deep water rich in nitrogen, phosphorous, and silica rises to take its place, producing thick blooms of diatoms. Diatoms, along with copepods and krill fattened on the extraordinarily high carbon concentration—34.2 ounces (970 g) of carbon per 10.8 square feet (1 m^2) of water per year—feed vast shoals of anchovies. Energy moves continuously through the food web; anchovies sustain seals and seabirds. The cool, upwelling waters of the Humboldt Current are bountiful, yielding millions of tons of anchovies every year. At one time, anchovies fed millions of cormorants, boobies, and pelicans living on the Chincha Islands and other rocky outcrops off coastal Peru. There the bird droppings were piled high into mountains of guano. This phosphate-rich fertilizer, mined and exported to Europe, was so prized in the nineteenth century that at one time it guaranteed Peru's foreign debt.

Along other coasts, other cool, nutrient-laden currents support thick forests of kelp, dense schools of fish, and feeding frenzies. Each year off the coast of South Africa, dolphins, sharks, seals, and gannets swoop in on shoaling sardines, who, in a desperate attempt to escape, fling themselves to certain death onshore. Off the coast of California, as fog washes in over San Francisco in the spring, cold upwelling water fuels a diatom bloom. The copepods and krill that feast in this meadow give life to a many-layered food web encompassing sardines, Dungeness crabs, seals and sea lions, humpback whales, and thousands of seabirds: puffins, petrels, murres, and cormorants.

Throughout the sea, the living nourish each other and the dead give sustenance to the living, generation after generation. The fundamental elements of life cycle and recycle from drifting plants to swimming animals and back into the sea. Nothing is lost. What sinks to the seafloor eventually, after millions of years, turns to stone and is raised into the rocks of high mountains. Wind and rain wear away their grandeur and wash them into the sea, where, once again, their constituent elements may give life to drifting plants.

The chalk cliffs of France and England, built from the skeletons of drifting marine plants, are now eroding and returning to the ocean. A rushing tide may carry the dissolved rock out to the North Sea to weave the web anew, provisioning blooms of floating plants, swarms of copepods, schools of herring, and schools of cod, each in its turn. The phosphorous, iron, and carbon that keep earth's ocean green today have been recycled many times. These building blocks may have fertilized blooms that fed the sea's first fish.

Arrival of Fishes

Some 370 million years ago, a fish crawled out of the water. Its fins were more like limbs, and it had lungs as well as gills. It moved back and forth between land and the coastal estuary, lagoon, or river from which it came, but eventually stayed ashore, beginning a life on dry land. Over millions of years, its descendants evolved into frogs and salamanders, dinosaurs, birds, and snakes, and eventually into monkeys, apes, and humans. Though the fish of the sea seem a world apart, and though our lives have substantially diverged from theirs, our origins are oceanic. Their lives gave rise to ours, and what they gave us is fundamental to our existence.

Origins

Fish evolved in the Devonian ocean, some 370 million years ago. Their history is beautifully and abundantly preserved in a fossil-filled cliff at Miguasha National Park (background), on the south side of Canada's Gaspé Peninsula.

To fish we owe our backbones. "Walking fish" come from a long lineage of vertebrates. Fish with notochords, a stiff rod of tissue that constituted the earliest "backbone," appeared in the Early Cambrian ocean. It was an inconspicuous start. In a sea teeming with trilobites large and small that crawled, burrowed, and swam, that actively hunted prey and were well equipped to hide from predators, these ancestral fish— no larger than a paper clip—didn't stand out. Yet as millions of years passed, trilobites would lose their marine hegemony and become extinct, and fish would rise to the top of marine food webs.

Cambrian waters were their humble beginnings. Several hundred fossils of the first fish have been found in the area around Lake Dianchi, near Chengjiang, China. *Haikouichthys ercaicunensis* not only had a notochord (which in later fish would evolve into a spinal cord and backbone), it had a head with a brain and a pair of eyes, incipient nostrils, the beginnings of ears, a heart, gills, and a sail-like dorsal fin. It had a mouth but lacked jaws. *H. ercaicunensis* lived approximately 530 million years ago. It would become extinct, but its descendants would flourish.

Remains of jawless fish from 450 million years ago have been found in arid Bolivia and Colorado, in places that long ago rested at the equator and were covered in saltwater. Jawless fish reached their peak during the Devonian period between 355 and 410 million years ago. They, like many species in the sea's long lineages of fish, evolved, rose to prominence, and then declined, many to become extinct. What we know of those whose lineages have died out we know from their fossils.

Some fish still walk on the seafloor today, using their pectoral fins. Above left: Shortnose batfish (Ogcocephalus nasutus), a Caribbean ray-finned fish. Other surviving ray-fins include Ziebell's handfish (Sympterichthys sp.), above right, Tasmania, and below, the red handfish (Sympterichthys politus), Tasmania.

Colliding continents and closing ocean basins create a book of life, a record of evolution. On the south side of Québec's Gaspé Peninsula, on the shore of the Restigouche estuary, lies Miguasha National Park. There, where the river meets the bay, a fossil-filled cliff rises from the beach. The cliff holds a remarkable picture of life in the Late Devonian, some 370 million years ago, a time known as the Age of Fishes. Then, sand and silt and mud that built the cliff lay farther south, in a warm, tropical estuary that teemed with fish. Earth's continents had begun drifting together: North America had collided with Scandinavia and Russia, and with Avalonia, whose lands would later drift apart and rest in New England and Newfoundland, Scotland, and Morocco. The collisions raised towering mountains that were immediately eroded.

Wind and rain turned their peaks to sand and silt, which rushing rivers carried down toward the sea. What is now Miguasha lay at the edge of that continent, in the tidal reach of a river, just as it does today. Grit that made the mountains washed into the river, burying the ancient estuary. Today continuing erosion removes particles that once made mountains, "exhuming" the ancient estuary and exposing fossils of those that once lived there.

The cliff at Miguasha holds fossils from almost every group of fish that evolved in ancient seas. A few jawless fish lived in the old estuary, filtering their food from the water or feeding along the bottom. Some were narrow, eel-like fish with downturned tails; others had bony head shields and bodies covered with scales. They were among the few jawless fish that had not yet become extinct, but their time, too, would end. As ocean basins that came before the Atlantic narrowed and lands that would eventually unite as one continent (Pangaea) began to coalesce, the shallow continental shelf, where the sea's early fish lived, began to shrink. Competition in this diminished world was intense; many jawless fish died out. Today their only living descendants are freshwater lampreys and deep-sea hagfish. Miguasha records not only those who departed, but those who came next. The successors, too, are entombed in the seaside cliff.

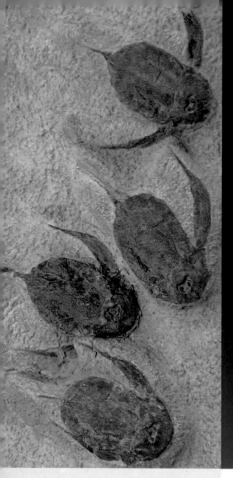

Jaws

A set of jaws may have made a difference between who lived and who died. When one of the bony arches that support fish gills evolved into jaws, the arms race between the sea's predators and prey escalated. Many jawless fish came up short. Fish with jaws and teeth can capture, crunch, and grind their prey. In addition, improved hearing and sensitivity to vibrations in the water aided them in recognizing danger and locating meals. An early jawed fish, abundant at Miguasha, was the short-lived but widespread, numerous, and bizarre placoderm *Bothriolepis*. It had the tail of a fish, but its front, covered with plates of bony armor, resembled a small, boxy turtle. It scavenged the nutrient-rich mud on the seafloor, aided by stiltlike pectoral fins. This design, however odd, was wildly successful. *Bothriolepis* spread from shallow coastal waters up into freshwater rivers on every continent.

One of the largest fish in the Devonian world, the giant placoderm *Dunkleosteus*, lived in the shallow waters of a sea that covered much of what is now the continental United States. Twenty feet (6 m) long, *Dunkleosteus* had a huge, bony-plated skull, powerful jaw muscles, and jawbones with self-sharpening edges that could slice through and dismember prey. Fierce and aggressive, it was the ocean's top predator, opening its gaping mouth in a split second, then snapping it shut with enough force to break bone.

Placoderms were not the only predators evolving in Devonian waters. Across the Restigouche estuary from Miguasha Point, in New Brunswick, is a site containing much older Devonian fossils. The site, like Miguasha, was once an ancient estuary near an ancient shoreline of North America. It too was buried by eroding mountains and then uncovered, enabling paleontologists excavating at low tide to find fossils in the riverbank. One fish, known to paleontologists only by the teeth it left behind, had been named *Doliodus problematicus*, the "problematic deceiver"; scientists suspected it was a shark, but no one knew its identity for certain. A fossil in the Restigouche riverbank solved the mystery. It was indeed a shark, found amid the remains of 3-foot (1 m) -long giant sea scorpions it may have been eating. It had pectoral fins with spines and jaws armed with sixty working teeth. Over 400 million years old, it is, so far, the oldest intact shark fossil paleontologists have discovered.

Sharks would endure as the sea's top predators long after placoderms made their exit. Some, like *Helicoprion*, had dinner-plate-sized coils of serrated teeth. Without a fossil of the lower jaw itself, it's hard to imagine how, or if, this shark closed its mouth. For many other sharks, the body plan that evolved in the Devonian stayed essentially the same. The sleek, spindle-shaped body; the keen sense of smell; the snout equipped with the ampullae of Lorenzini, jelly-filled receptors that pick up electrical signals generated by the muscle activity of potential prey, even when it is buried in the sand; and the sharp teeth that replace themselves every few weeks—all made an exquisite hunter that quickly populated deep and shallow waters throughout the ocean.

Sharks evolved in the Devonian ocean. Below: The dinner-plate-sized coils of the now extinct Helicoprion.

Bones

Sharks were fast from the beginning, chasing fish, catching them by their tails, and swallowing them whole. They preyed on ray-finned fish, fish with skeletons of bone and buoyant, air-filled swim bladders. Before the evolution of swim bladders, fish used their fins to keep from sinking. With swim bladders to keep them afloat, the agility of fish improved, and they began using their thin, delicate webbed fins to steer and brake.

Early ray-fins, whose remains are preserved in the cliff at Miguasha Point, were harbingers of a world to come. Like the small vertebrates in the large world of Cambrian animals, ray-fins were a minority in the Devonian ocean, small fish among the placoderms. From few, though, came many. Their descendants would become by far the most diverse and abundant of earth's vertebrates. They would proliferate throughout the water world, spreading through the sea and swimming upstream to make their homes in freshwater rivers and lakes. They would live in the freezing Arctic and the warmth of tropical reefs, in cold, clear mountain streams, and steamy, weed-choked swamps. They would become small herring and giant tuna, solitary anglerfish resting in dark and barren depths, and schooling cod at the sea's fertile surface. Their species are richly diverse, and their numbers are great. By almost every measure of evolutionary success, they triumphed.

We did not descend from ray-fins. Our lineage goes back not to them but to lobe-fins, fish with fleshy fins. These were the animals that walked from the sea and ultimately gave rise to mammals. Terrestrial lines of human evolution—from amphibians to tiny shrewlike mammals, to apes, and then to humans—did not begin at the edge of the sea. Our origins are oceanic, and we are highly specialized fish.

While lobed-fin fish thrived during the Devonian very few are alive today. Among our few living marine relatives are fleshy-finned coelacanths, once believed extinct. They still turn up in fishing nets in the Indian Ocean off Madagascar and in coastal waters off Sulawesi, Indonesia.

Ray-finned fish were a minority in the Devonian ocean, but today they are by far the most diverse and abundant of earth's vertebrates. Below and opposite, left to right: Yellow moray (Gymnothorax prasinus), *New South Wales, Australia; French angelfish* (Pomacanthus paru), *Saint Lucia, Caribbean Sea; great barracuda* (Sphyraena barracuda), *Belize, Caribbean Sea; ribbon moray* (Rhinomuraena quaesita), *Sangeang, Indonesia; longfin bannerfish* (Heniochus acuminatus), *Solomon Islands; and male mandarinfish* (Synchiropus splendidus), *Malapascua Island, Cebu, Philippines.*

Opposite: A whitemargin stargazer (Uranoscopus sulphureus) *awaits its prey in the sea bottom off the coast of Papua New Guinea.*

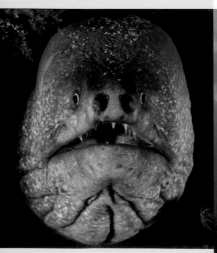

The broad leaves of Archaeopteris, *one of earth's first trees, fell into the water, helping to create vegetation-filled shallows where fish would learn to walk.*

*Opposite: Two-spot New Zealand demoiselles (*Chromis dispilus*) school in the Poor Knights Islands Marine Reserve, New Zealand.*

CONFIGURATION OF CONTINENTS
EARLY DEVONIAN
390 MILLON YEARS AGO
THE AGE OF FISHES

Ancient Landmass

Modern Landmass

Subduction Zone
(triangles point in the direction
of subduction)

Seafloor Spreading Ridge

Drifting continents and seafloor record evolution and make it possible. Streams and rushing rivers carried rock and silt down from mountains raised as continents collided, building vast floodplains at the edge of the sea. Somewhere in one of those floodplains, small plants made a life in the mud, piping water through tiny, branching stems and breathing through tiny pores. The greening of continents was under way.

Fossils of these green pioneers have been found in the Restigouche estuary where sharks and giant sea scorpions once swam. Millions of years later, when the Miguasha fish lived there, the riverbank had changed. By then plants had roots and leaves, and although the birches and trembling aspens that grace the nearby woodlands today had yet to evolve, there were stands of 23-foot (7 m) *Archaeopteris*, one of earth's first trees. Its broad leaves shaded the riverbank, and its extensive roots helped weather the rock into thick soil. As the seasons changed, its leaves and branches fell into the water. Some would be buried and become fossils; others decayed.

Disintegrating mountains and homesteading plants created shallow, vegetation-filled water. Fish moved in. No one knows why. Perhaps they sought smaller prey hidden in the leaves. Perhaps they fled from large predators in the deeper water. Perhaps they were squeezed by increasing competition for food and shelter as earth's continents edged closer together and the continental shelf shrunk. However they got there, in their new habitat fish learned to walk.

They walked so they could breathe. The warm, shallow water held less oxygen. *Tiktaalik* couldn't stand in the open air, but it could prop itself up on its fins, flex its neck, raise its snout out of the water, and take a gulp of air. What worked for one purpose would eventually work for another. When the time came to walk from water, the descendants of *Tiktaalik* and *Acanthostega* would be well prepared.

That time eventually came. Perhaps some fish left shallow, vegetation-choked waters of their own volition, their young finding an abundance of insects in the meadows and woodlands onshore, then staying on in a rich, unclaimed habitat free of predators. Perhaps they left from necessity. Plants that turned a dry and dusty land green brought calamity to the ocean. As they breathed and removed carbon dioxide from the air, land plants helped cooled the earth, setting off a glaciation that caused the sea to recede and marine homes to dry up. In addition, as fallen leaves and branches rotted in estuaries and swamps, the decay used up oxygen, and fish began to suffocate. Whatever prompted the exodus, those that climbed ashore and survived became our ancestors.

Many that stayed in the sea died, unable to manage in the stagnant water. The Devonian saw the rise of fishes, the greening of continents, the ascent of fish onto land, and a mass extinction of those dwelling in the sea. It was not the first time—nor would it be the last—that species would rise and fall on the promise and catastrophe created by the complicated partnership between the restless earth and the life it created.

Siberia

North China

Caledonide Mts.

Kazakhstania

South China

Malaya

EURAMERICA
(Laurentia & Baltica)

Southern Europe

Australia

Northern Appalachians

Arabia

India

Antarctica

RHEIC OCEAN

GONDWANA

Africa

South America

5 Climate from the Sea

Wind and ocean currents distribute the sun's heat as it falls on the earth, making the planet hospitable to the life we know.

Blazing sun, atmosphere, ocean, drifting continents that raise mountains and block wind, the path of earth's orbit around the sun, and the tilt of its axis—all give earth its unique climate. Sunlight falls unevenly on the planet: without the sea, the tropics would scorch, and the ends of the earth would plunge into a deep freeze. Ocean currents, coupled with wind, mute these extremes, spreading heat from the tropics to warm higher latitudes, making the planet hospitable to the life we know. The distribution of heat is uneven, and inconstant. When currents shift, climate responds. Living on shore, we may not see the shift, but we feel its effects. Stirrings in the sea touch continents, soaking land with life-giving rain or leaving it arid and thirsty, making the difference between feast and famine, life and death. Sea changes have altered the course of evolution, brought about the rise and fall of cities and civilizations. The sea is our lifeline; our well-being is tied to its rhythms.

El Niño: A Natural Rhythm in the Sea

A pool of warm ocean water, the warmest in the world, lies in the western Pacific along the equator in the sea off Indonesia and the Philippines. Its size approaches that of the continental United States, and its temperature can reach as much as 86 degrees Fahrenheit (30°C). Every few years the trade winds slacken, and the pool drifts east across the sea, appearing around Christmas in coastal Peru, where it is named El Niño, "the Child."

This Pacific warm-water pool cycling back and forth across the ocean has been a natural rhythm of the sea for the last 130,000 years, bringing rain and taking it away, suppressing cold, fertile upwelling water in the Humboldt-Peru Current and then releasing it, taking and then renewing life. Dense schools of anchoveta school in the cold, nutrient-rich waters of the current. When El Niño appears, a rich food web collapses. Copepod and krill populations crash, anchoveta stocks plummet, and millions of cormorants, pelicans, and boobies starve. While some animals suffer, others thrive: shrimp, scallops, sharks, and tuna move into the warming waters off Ecuador, arriving and departing with El Niño.

El Niño resonates in the sea and on land, a tangible influence on earth's climate. The migration of the warm pool during an El Niño leaves Indonesia, Australia, and other Pacific islands plagued by drought. In arid Peru, rains fed by warm seawater are both blessing and curse, bringing life to the desert, recharging rivers and groundwater, while releasing catastrophic floods that destroy houses, roads, and bridges, and contaminate drinking water. El Niño has brought heavy rains and mudslides to California and typhoons to the Marshall Islands, while reducing the punch of Caribbean hurricanes. It even triggers infectious disease in humans.

El Niño brings rain and takes it away. Above: A skeleton of a sheep (Ovis aries) that starved to death on drought-affected land, Cunnamulla, Queensland, Australia. Right: Deer mice (Peromyscus maniculatus) trigger the rise of human hantavirus infections when El Niño comes to the American Southwest and the population of mice rises.

El Niño can bring drought to Australia, drying Lake Hume (opposite above), while watering the woodlands of the American Southwest. Opposite below: Junipers and pinyon pine growing on the Glorieta Mesa near Santa Fe, New Mexico.

In the 1997–98 El Niño, 22,000 people died, mostly from outbreaks of infectious disease triggered by a warming sea. Bacteria blossomed in Peru's coastal waters, setting off a deadly cholera epidemic. In its wake, El Niño has also brought cholera to Bangladesh and increased malaria to Colombia, Peru, and Venezuela. Its rhythms have reached far inland, igniting the rodent-borne hantavirus in the southwestern United States.

In the arid Four Corners region of New Mexico, Arizona, Colorado, and Utah, El Niño's rains water desert grasses and dry woodlands of piñon and juniper. The effects cascade through the ecosystem. With twice as much rain, the desert blooms, producing a cornucopia of seeds, berries, piñon nuts, and grasshoppers. Deer mice feast on the bounty, and their numbers swell, exposing people coming into contact with mouse droppings to a serious and sometimes deadly lung infection. When El Niño wanes, the rains disappear, grasses and woodlands dry, rodent populations dwindle, and cases of Sin Nombre decline, an elegant illustration of how the rhythms of the sea reach inland.

The Rise of Agriculture

Agriculture is the foundation of civilization. A transition from hunting and gathering to farming and raising animals ultimately gave rise to the concentration of large numbers of people in cities, the division of labor, the production of pottery, the rise of trade, the development of writing, and the establishment of armies and states. In the Fertile Crescent, on the banks of the Euphrates River, that transition began as the last ice age waned and an Atlantic current was suppressed.

As the ice receded, the Fertile Crescent began to warm. Rivers and streams flowed, woodlands of oak and pistachio trees appeared, and wild wheat and rye grew on the plains. In this warming world, hunter-gatherers didn't have to travel far for food. From their homes along the Euphrates River, they could easily collect wild cereal grasses in the spring and nuts in the fall, and hunt herds of gazelle running on the plains.

The last ice age ended not smoothly and continuously, but spasmodically. Roughly 12,900 years ago, the gentle warming, which had produced plentiful food in the Tigris-Euphrates valley, was interrupted. Across the ocean on another continent, melting ice had pooled into Lake Agassiz, a large lake reaching from Minnesota and North Dakota up into Manitoba and Saskatchewan. An ice dam plugging the Gulf of St. Lawrence burst, draining the lake into the North Atlantic and flooding the sea with fresh water. The water, too buoyant to sink into the ocean's cold, deep currents, floated at the surface, slowing the sea's circulation, weakening the Gulf Stream. The current carried less heat to the poles, triggering a cold snap that would last more than one thousand years.

A change in climate wrought by the sea helped bring about the rise of agriculture, leading to the cultivation of humanity's major food crops. Among the first was wheat.

The Natufians, hunter-gatherers in the Fertile Crescent, felt the chill. During this period, known as the Younger Dryas (named for a polar flower), the distant ocean cooled, drying their land. Nut forests retreated and wild cereal grasses dwindled. The villagers adapted to the ensuing food shortages by cultivating grains and domesticating sheep and goats. The ruins of an ancient Syrian village, Abu Hureyra, now lie beneath the waters of Lake Assad. The village, excavated before the dam was built, records a pivotal moment in human history when foragers, responding to a change in climate wrought by the sea, became farmers.

Below: Two views of Mount Pinatubo, a volcano born of descending seafloor, which has left its signature on earth's climate.

A Closing Sea Leaves Its Fingerprint on Earth's Climate

In 1815 an Indonesian volcano, Tambora, erupted explosively and violently. As the Pacific narrowed and pieces of its seafloor descended back into the earth, Tambora hurled millions of tons of ash and sulfur into the atmosphere. The sulfur formed an aerosol mist that scattered and absorbed incoming sunlight, sending the earth into a chill felt around the world. Snow fell in New England in July, crop failures were widespread, and 90,000 people across the world died from starvation and disease in 1816, the "year without a summer."

Volcanoes born of descending seafloor still leave their signatures on earth's climate. In 1991 Mount Pinatubo in the Philippines shot 10 million tons of sulfur into the atmosphere, cooling the earth nearly one degree Fahrenheit (half a degree Celsius), thus temporarily buffering the planet from the effects of human-induced global warming.

The Collapse of the Maya

A reduction in rains sent by the sea caused drought that may have contributed to the collapse of the great Mayan city of Tikal.

Gentle changes in the temperature of the sea and the strength of the wind can alter the course of civilizations. Tikal, Copán, and Palenque were once large and glorious city-states of the Classic Maya civilization. Their rulers built sophisticated astronomical observatories, temples, pyramids, and ball courts; recorded their histories in glyphs engraved in stone monuments; and made elegant carvings from wood and jade. Theirs was a maize-based agriculture, and they supported it by turning their quarries to reservoirs to catch the rain, and building extensive gravity-fed irrigation canals. The rain itself came from the Atlantic, where trade winds blowing from either side of the equator meet and warm, moisture-laden air rises from the water.

Before the close of the first millennium A.D., the great cities of the Maya were abandoned, and this long-lived and powerful civilization collapsed. Warfare among the kings, deforestation, and soil erosion may not fully account for the fall of this wealthy and widespread civilization. An additional explanation may lie in the ocean off the coast of Venezuela, where layers of mud built into the seafloor contain debris washed down hundreds of years ago by rivers. The layers tell of times when the area's rivers were flooded with rain and times when the rains had dissipated. They suggest that when the Maya civilization collapsed, its cities had been besieged by severe and recurring drought. Currents and sea surface temperatures in the Atlantic had shifted, rain stayed to the south, and water evaporated from the sea no longer rained on Mayan maize.

Feast and Famine in the Fertile Crescent

The sea gives life but also takes it away. Between the Tigris and Euphrates rivers, on the plains of Habur near the Iraq-Syria border, are the remains of a once vibrant city, Tell Leilan. Tell Leilan was a provincial capital and agricultural center of the prosperous Akkadian Empire, which extended along the rivers from their headwaters to the Gulf of Oman. Extensive cultivation of grain supported its prosperity: the Akkadians grew barley on the fertile rain-fed floodplains between the rivers. Archaeologists excavating the old city uncovered residences and a schoolroom, the palace and administrative building, cuneiform tablets, and the city gate. They also uncovered a thick layer of dust without any sign of human presence.

About 4,200 years ago, most of Tell Leilan was abruptly abandoned. Archaeologists suspected drought. Dust buried in the seafloor of the Gulf of Oman supplied the evidence: blown off the Mesopotamian floodplains, it marked the onset of a severe drought that lasted 300 years and brought the Akkadian Empire to its end.

The sea provided a record of rains withheld. Those rains, like those that nourished the Maya civilization, come from the sea. The plains of Habur are watered by rain that fills the Tigris and Euphrates rivers, rain that is born in moisture-laden air rising from the Atlantic and the Mediterranean. Distant stirrings in the Labrador Sea summon those rains. When, periodically, that sea chills slightly more than usual, the winds strengthen and the eastern Atlantic warms. Rain falls to the north, lowering water flow in the Tigris and Euphrates rivers by as much as 40 percent. A sea change in the Atlantic 4,200 years ago meant a civilization could no longer feed itself.

The Greening, Then Drying of the Desert

During the dry season, a harsh, dusty wind, the Harmattan, blows through the Sahara and the grass plains of the Sahel at the desert's southern edge. Baobab and acacia grow in the Sahel, where nomadic herdsmen tend their cattle, sheep, goats, and camels, and farmers cultivate fields of sorghum and millet. They all await the rain, which arrives from the sea on warm, moist monsoon winds, and whose scarcity or abundance depends on fluctuations in the temperature of the Atlantic. Today the rains are unreliable and often scanty, as they were during the last ice age, when the Sahara was even larger than it is today. The Sahara, earth's largest desert, has not always been so dry.

About 10,500 years ago, as the ice receded and the earth continued to warm, African monsoon rains shifted north, turning the desert green. Rain pooled into lakes, and rivers flowed. The barren dunes blossomed with life; animals roamed plains lush with wild cereals. People moved in. Following the rain, they migrated into the Great Sand Sea, now part of Libya, eventually becoming livestock herders, tending domestic sheep, goats, and cattle. The rains lasted for over three thousand years.

Locust on a millet plant in Mauritania. The scarcity or abundance of water for food crops in the arid Sahel depends on rain blown in from the sea.

The Sahara has not always been as dry as it is today. In the past, the rise and fall of human occupation in the Sahara have followed the advance and retreat of rain given by the sea. Above: Western Desert, Egypt. Opposite: A sandstorm between Douentza and Timbuktu, Mali.

Then temperatures in the sea changed, and the monsoons shifted south. The pastures returned to desert, and the herders retreated, taking their animals to oases at the edges of the sand and to the high plateau of Gilf Kebir, which still had water and wild grain. These areas, too, would return to sand: cave paintings at Gilf Kebir depict swimmers in a river that has long since dried. Water from this time still lies deep in the sand, and Libyans are now extracting it to irrigate the desert.

In parts of Africa the savanna is still turning to sand. During the Middle Ages, Timbuktu was a fabled city, a wealthy center where caravans bearing salt and figs from the north met gold and slave traders from the south, where wheat grew on the fertile flood-plain of the Niger River, and where thousands of scholars studied Islamic law, medicine, science, and mathematics at the city's great university. Today the centers of African trade have moved elsewhere, and Timbuktu is a quiet backwater. Advancing sand swirls around the great mosques of this former intellectual and spiritual capital, whose greatness is recorded in its trove of ancient and crumbling manuscripts.

Farther east, in Sudan, the monsoons are still weak. In the last sixty years, average rainfall at the northern edge of the Sahel has declined by 34 percent, turning millions of acres of already marginal pasture into desert in a country that already was mostly desert. As the desert advanced, herders with increasingly large flocks were forced south in search of water and pastures, where they fought with farmers over land. The scarcity and hardship caused by the advancing desert and the degradation of land exacerbates religious, ethnic, historical, and political struggles underlying the war in Darfur. Droughts have already caused devastating famines in Sudan. If more grassland continues to turn to sand, Sudan will become less and less able to produce food for its people. Over three-quarters of the world's population live in the tropics, where the strength of monsoon rains called by the sea helps determine whether there is food to eat.

Mass Extinction

Evolution proceeded slowly, even lethargically, in earth's early seas. For more than three billion years, or 85 percent of earth's history, the ocean was inhabited primarily by organisms too small to be seen with the naked eye. The quiet ended after single cells containing chlorophyll had pumped enough life-giving oxygen into atmosphere and ocean to sustain the metabolism of a large animal. Then a profusion of life burst forth. In a comparatively short forty million years, the sea filled with animals of almost every type of body plan known today: sponges and corals, anemones and jellyfish, worms and snails, fish and crustaceans. Finding many ways to make a living in their watery world, they grew in diversity. Lacy bryozoans packed dense colonies with thousands of animals. Sedentary sea lilies and shelly brachiopods filtered food from the surrounding seawater. Trilobites moved freely, swimming and crawling in the mud.

Few animal species from this time survive today. Trilobites once dominated earth's ocean; all described species, some 15,000, are extinct. Giant sea scorpions and boxy armored placoderms that once lived in Canada's Restigouche estuary are gone, as is the once fearsome predator *Anomalocaris* and the first walking fish. Earth's history is littered with extinction, and the seafloor holds the history of those that have disappeared. The record is sobering. Virtually all species, 99.99 percent, that ever lived on this planet are now extinct.

Strange and Wondrous Sea

Sea anemones evolved in the Cambrian ocean. Their descendants live on today. From the Channel Islands National Marine Sanctuary and National Park, California, anemone (Anthopleura elegantissima), top, and strawberry anemones (Corynactis californica), above.

Background: Brachiopod fossils (Camarotoechia sp.), Willow River Formation, Michigan.

In the first 300 million years after the Cambrian explosion, the sea bore little resemblance to the ocean of today. Today copepods pull food from the water with feathery antennae and burst away from predators at 200 body lengths per second. They are the ocean's predominant grazers, but their reign is recent. When trilobites roamed the seafloor, swarming colonies of graptolites floated near the surface. Each graptolite lived in a tiny tube, arranged in strands of 100 or more. Once widespread and abundant, their heyday has long passed. They are gone but not forgotten, their fossils preserved as sawlike zigzags in old seafloor now raised onto the edges of Newfoundland, the moors of Scotland, and the mountains of Utah. For nearly 100 million years, graptolites flourished, species after species arriving and departing, marking the divisions of geologic time.

Remnants of old seafloor hold fossils of animals now extinct. Above, left to right: Rudist bivalve (Hippurites radiosus), *Late Cretaceous, Charente, France;* brachiopod (Spirifer *sp.), Devonian, Sylvania, Ohio; crinoid* (Agaricocrinus splendens), *Mississippian, Montgomery County, Indiana; crinoid* (Uperocrinus nashvillae), *Mississippian, Iowa; brachiopod cluster* (Cyclocanthoria kingorum), *Permian, Hess Canyon, Texas.*

A family walking the shore of today's ocean at a very low tide, looking amid mud and rock for animals with two hinged shells, might find clams, mussels, or oysters. In the Paleozoic sea they would have found the now strange and unfamiliar brachiopod. Its shells came in many shapes and sizes: rounded like helmets, smooth, perforated, ridged like scallops, spiny, or winged. As adults, brachiopods lived sedentary lives, resting in large beds on the seafloor or attached by a stalk to the hard bottom, filtering food from water flowing by. Once they were ubiquitous. Their numbers now greatly diminished, they hang on, mostly in the remote icy waters of Antarctica and in the recesses of underwater caves.

Great stands of sea lilies blanketed the floor of the Paleozoic ocean. Though crinoids looked like flowers—with a stemlike stack of discs anchored to the seafloor and topped by a cup fringed with feathery arms—sea lilies were animals, filtering their food. They prospered, growing to different heights, feeding from different layers in the water column. Their arms evolved into a variety of shapes, adapting to varying strengths of surrounding currents and varying sizes of drifting food particles.

Right: Sea lilies (crinoids) once covered the seafloor. Though their numbers and species have decreased, a few still live in the contemporary ocean. Here, a feather star on a red sea fan, from the Solomon Islands.

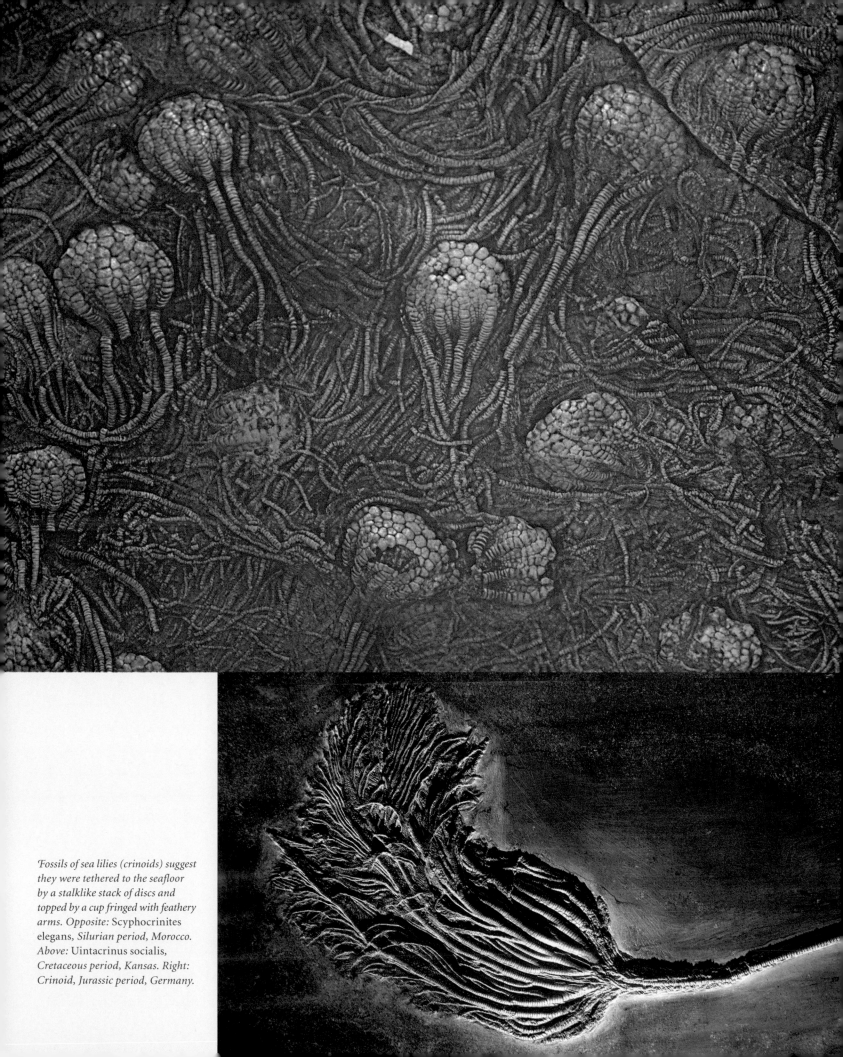

Fossils of sea lilies (crinoids) suggest they were tethered to the seafloor by a stalklike stack of discs and topped by a cup fringed with feathery arms. Opposite: Scyphocrinites elegans, Silurian period, Morocco. Above: Uintacrinus socialis, Cretaceous period, Kansas. Right: Crinoid, Jurassic period, Germany.

The sea that gave rise to graptolites, brachiopods, and sea lilies has long drained away, but vestiges of the old seafloor and its fossils remain, now part of continents. The gray, lonely cliffs of Green Point in Newfoundland's Gros Morne National Park are built of layer upon layer of seafloor, angled upward as the Iapetus Ocean, the Atlantic's predecessor, closed. The cliffs are etched with the remains of graptolites.

El Capitan, looming behind a salt basin pool in Guadalupe Mountains National Park, Texas, was once a vibrant reef in a shallow sea. Its inhabitants bore little resemblance to corals today.

Towering above the hot Texas desert, about 100 miles east of El Paso, is a steep limestone cliff, remnant of a vibrant reef that grew in a shallow sea once covering west Texas. El Capitan and the nearby reefs, rivaling Australia's Great Barrier Reef in richness, were home to crinoids and sea urchins, algae and trilobites, and over one thousand species of brachiopods. When the sea drained away, the reef was buried, but as the Pacific Ocean closed and its seafloor descended beneath western America, the reef was uplifted. Today it rises in stark relief, a monument to a distant and now unfamiliar world whose time suddenly ended.

Snuffed Out

When the end came, approximately 252 million years ago, it was devastating. The Great Dying, as the Permian mass extinction is often called, eliminated as many as 90 percent of all species that lived in the sea, far more than the death toll from the Cretaceous extinction 65 million years ago that killed the dinosaurs, or the three other mass extinctions that mark earth's history. The Meishan cliffs in southern China record the Permian annihilation. Their layers of limestone, formed in the same sea that housed El Capitan but on the opposite shore, are full of fossils: 333 species of brachiopods and cephalopods (relatives of today's squid), fish and trilobites, corals and lacy colonial bryozoans. Higher in the cliffs, the younger layers of limestone and clay are bare: 94 percent of the species had vanished.

Not only did species, genera, and families disappear; entire ecosystems collapsed. Early in earth's history, life consisted of single cells making carbon, first in the heat of deep-sea hot springs and then in the sea's sunlit surface. At some point, single cells engulfed each other and dwelled together. By means still debated, a cell evolved with its DNA contained in a nucleus. Later a nucleated cell swallowed a photosynthesizing bacterium; resisting digestion, it became the chloroplast of earth's first plant.

As plants and animals evolved together, their partnerships extended beyond individuals into entire communities whose members depended upon one another for survival. Energy moved through layered food webs containing tiny photosynthesizing cells, herbivorous grazers, and carnivorous predators. When they died, their waste supplied nutrients for the cycle to continue. Food webs gave rise to larger and longer-lived animals whose very existence hung on the work and health of others.

During the Permian extinction, the entire system—both individuals and the communities they built—fell apart. The rules for survival suddenly changed. Hegemony was no advantage. Those who had flourished disappeared. Evolution may indeed select for the fittest, but under severe duress, the definition of "fittest" may abruptly and unpredictably change. Whole forests of sea lilies, enormous coral reefs, and vast beds of brachiopods died. These sedentary, filter-feeding animals with heavily calcified shells and slow metabolisms were wiped out. Worms and mollusks that could dig into the seabed and fish that could swim away fared better. In addition, their faster metabolisms and more sophisticated circulatory systems were better equipped to expel the poisoned seawater. The poison was a common enough gas: carbon dioxide.

A Disaster of Earthly Origin

The Permian catastrophe probably began with a volcanic eruption more immense than any humanity has witnessed. An Iceland volcano, Laki, sits on the Mid-Atlantic Ridge, widening the Atlantic. Its history hints at the scale of the earlier disaster. In 1783 lava poured from Laki, swallowing farmhouses and churches and burning fields. Acrid fumes poisoned air and water, killing thousands of people, cattle, and sheep and withering crops. Pollution and famine induced by the eruptions killed 25 percent of Iceland's population. An acid haze hung over Europe, polluted the air, and killed more people.

Laki erupted for eight months; eruptions in Siberia 252 million years ago may have lasted for one million years. Today an area between the Lena and Ob rivers, between the Arctic Sea and Lake Baikal—an area about the size of the continental United States—is flooded with lava, in some places 4 miles (6.5 km) thick. The eruptions burned through some of the Tungusskaya coal, the world's largest coal basin, vaporizing it into carbon dioxide and methane, altering the composition of the atmosphere, and making it impossible for many of earth's inhabitants to breathe.

Today 20 percent of earth's atmosphere is oxygen. During the Permian extinction, oxygen levels dropped to 12 percent. In the sea, carbon dioxide and methane released from flowing lava and burning coal warmed the water, reducing its capacity to hold oxygen. As the earth warmed, temperature differences between tropics and poles decreased, slowing the currents, delivering less oxygen to the depths, favoring the proliferation of bacteria producing lethal hydrogen sulfide and carbon dioxide. Finally, as atmospheric levels of carbon dioxide rose, so did carbon dioxide concentrations in the sea, increasing the acidity of the water.

The combination—too little oxygen, too much carbon dioxide and hydrogen sulfide, and increasing acidity—was disastrous. Brachiopods, corals, and sea lilies were poisoned by hydrogen sulfide and carbon dioxide. They suffocated from insufficient oxygen, and their protective limestone shells dissolved in the acidic water. The tragedy is written into limestone from Permian ocean seafloor. The old seafloor, showing signs of dissolution from the time of the extinction, still stands, exposed in the steep mountains of southern China, in pastures high in the Taurus mountains in Taskent, Turkey, and in densely wooded hillsides of Kyushu, Japan. For some, death was instantaneous. For others, it was delayed; in waters with chronic low oxygen levels, animals grew more slowly and their reproduction was impaired, diminishing their numbers slowly but surely. A steady 1 percent population loss each year is enough to push a species to extinction.

CONFIGURATION OF CONTINENTS
LATE PERMIAN
255 MILLON YEARS AGO

- Ancient Landmass
- *Modern Landmass*
- Subduction Zone (triangles point in the direction of subduction)
- Seafloor Spreading Ridge

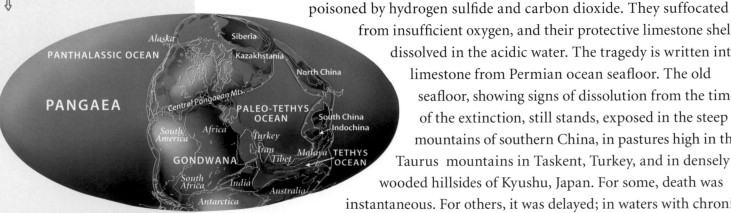

From Few, Many

Two species of sea urchin, including Miocidaris heschlerii, *bottom, from the Middle Triassic, Switzerland, survived the Permian extinction and gave rise to urchins (red sea urchin,* Strongylocentrotus franciscanus, *below, Channel Islands National Marine Sanctuary and National Park, California) and western sand dollars (*Dendraster excentricus, *bottom right, in Monterey Bay, California) living in the sea today.*

Bacteria survived. Sulfur-loving bacteria thrived, and bacteria that once covered the seafloor with stromatolite reefs and domes began building again, their remains scattered today in Greenland, Turkey, and Utah. These bacteria may be limited in size and evolutionary capacity—many of today's marine bacteria are virtually indistinguishable from their billion-year-old ancestors—but they are seemingly extinction-proof. Their versatility and flexibility in extreme conditions have enabled them to endure not only the Permian mass extinction, but others as well.

For the few larger organisms that survived the extinction, biology was destiny. The clams *Claraia*, *Eumorphotis*, and *Promyalina* tolerated low oxygen levels, reproduced rapidly, and could bury themselves in the seafloor. They flourished in the extinction's long and dismal aftermath, as did one brachiopod, *Lingula*, whose phosphatic shell resisted dissolution in the acidic water. Though the fossil record that followed the annihilation was sparse, these "disaster taxa" appeared again and again, like weeds. They left thousands upon thousands of shells on seafloor now exposed in the red desert rocks of Utah, the Italian Dolomites, and the sea cliffs of northern Honshu, Japan.

A few snails survived. Surviving snails, brachiopods, and bivalves were smaller versions of their predecessors: a shortage of food and oxygen caused this "Lilliput effect." *Lingula* is the size of a large seed, and the surviving snails the width of a paper clip.

Life is tenacious. A few survivors repopulated the ocean. A sea anemone grew a skeleton, giving rise to today's corals. From only two species of sea urchin evolved the multitude of flat sand dollars, plump sea biscuits, and purple and spiny sea urchins that thrive in the ocean today.

Legacies

Mass extinctions erased life but not its record. The airless waters that triggered mass extinctions also kept earth's biography. Ammonites, distant relatives of octopus and squid, swam in the Cretaceous ocean, maintaining buoyancy by filling their chambered shells with gas. Large marine reptiles, plesiosaurs and mosasaurs, preyed upon them. All disappeared in the Cretaceous extinction, never to return, but the seafloor retained the history of their lives. Some of those that died came to rest in water that held no oxygen. Spared the ravages of decay and decomposition, they achieved their own immortality and are among earth's few inhabitants whose remains are preserved in the fossil record. Coiled ammonites and giant marine reptiles up to 50 feet (15 m) long—animals that became extinct 65 million years ago—have been exquisitely preserved in anoxic mud that hardened into rock.

Marine plants and bacteria that escaped decay but did not become part of the fossil record left a different kind of legacy. Cell by cell, they accumulated on a sea bottom that lacked oxygen. Millions of years passed, and the rain of organisms built up the seafloor, layer by layer. The bottom layers of undecayed plant and bacterial remains sank, and in the heat of the depths they slowly cooked, turning to natural gas and petroleum. The light oil and gas flowed into reservoirs of porous rock like El Capitan and the other ancient limestone reefs of Texas, where they now supply the United States with oil and natural gas.

Pieces of seafloor, now exposed on land, contain the fossil record of large marine reptiles that are now extinct. Below: Plesiosaur (Plesiosaurus macrocephalus), Jurassic period, England.

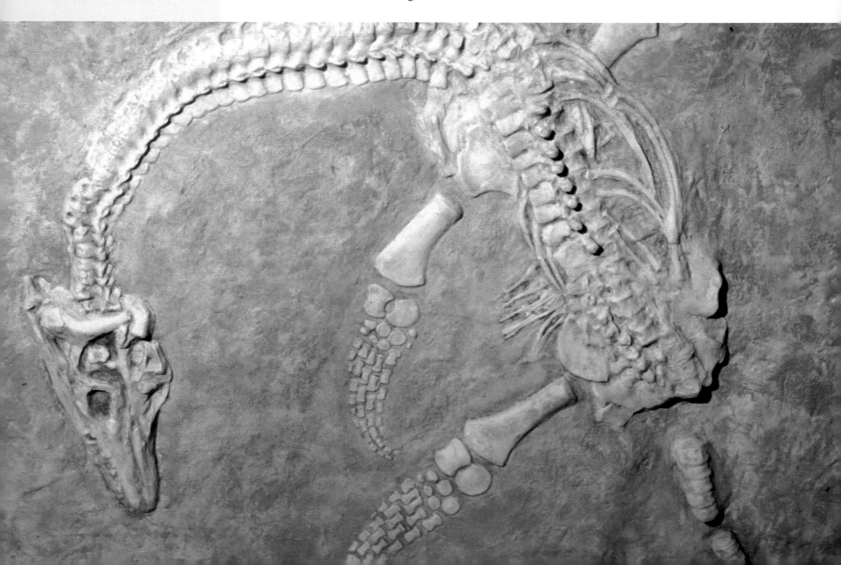

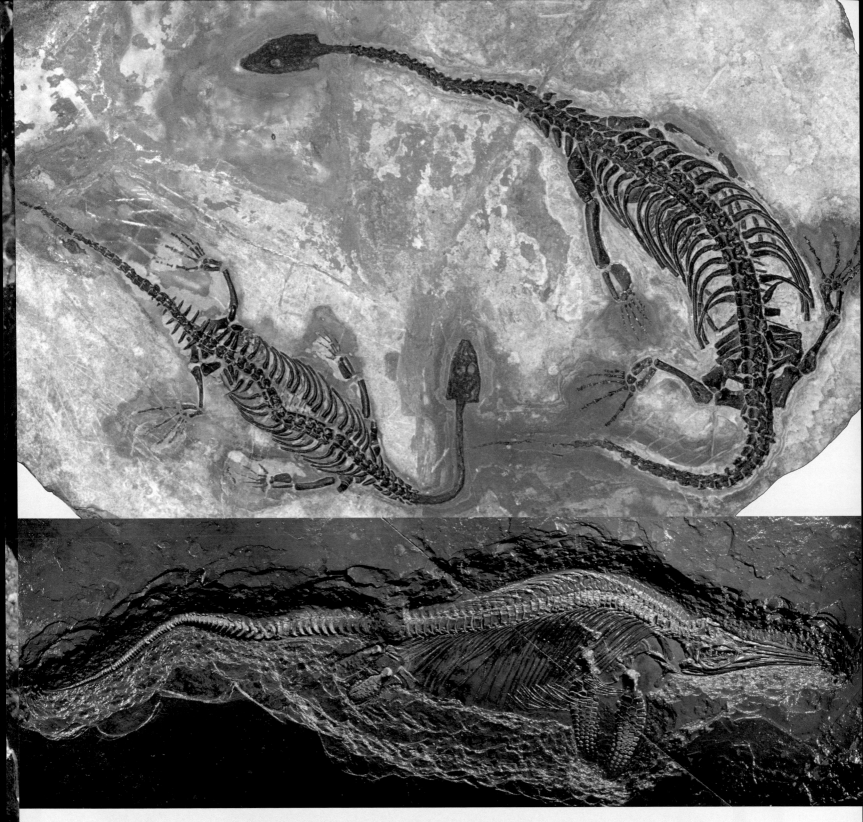

Top: Nothosaur (Keichousaurus hui), Triassic period, China. Above: Ichthyosaur.

Closing ocean basins bequeathed not only oil and natural gas but also an essential preservative, precious ores, and stone building material: layer upon layer of salt left behind as disappearing seas evaporated in the hot sun; shale and limestone compressed and baked into veins of marble and slate as seafloor was pressed into continents; gold and copper precipitated in deep-sea hot springs fueled by seafloor descending back into the earth. The closing of the Pacific Ocean has built one of the world's largest gold deposits at Yanacocha, Peru, and one of the youngest on the other side of the sea on Lihir Island in Papua New Guinea.

Kashmir's Dal Lake in the Indian Himalaya. Colliding continents and a closing ocean basin—forces of a restless earth—built the high Himalaya. Now humanity too has become a powerful force of nature.

The Anthropocene

Only one species in the genus Homo remains. On this last human species, the health of the earth depends.

distance between woodlands, supplementing their fruit diet with nuts and tubers, and turning the volcanic rock produced by the emerging sea into stone tools. They began to walk upright, and their brains doubled in size.

In the beginning, the human family tree had many branches. *Sahelanthropus tchadensis*, also known as "Toumaï," the "hope of life," came from the grasslands of Chad; "Lucy" and her younger sister "Salem" dwelled in the well-watered woodlands of Afar. *Homo habilis*, the "handy" man, from Tanzania's Olduvai Gorge, used stone tools to cut animal meat from bone. *Australopithecus garhi* from the Awash River used stone tools to butcher animals and extract marrow from the bones. *Homo ergaster*, the "workman," had long legs in a modern proportion. Upright *Homo erectus* coexisted at Lake Turkana with *Homo habilis* for half a million years and also migrated out of Africa to Java and China. Large-brained Neanderthals lived in the caves of Europe and ranged into Siberia, coexisting with modern humans on both continents.

From so many ancestors, only one lineage remains. In the genus *Homo*, all species are now extinct except *Homo sapiens*, the "wise man." The last human, a highly powerful and dangerous competitor, evolved in a flash of geologic time. In less than 200,000 years, *Homo sapiens* dispersed throughout the world, migrating out of Africa into Asia, west into Europe along the shores of the Black Sea, and east, following mammoth and bison over the steppes across the Bering land bridge into North America. In what are but a few moments of geologic time, modern humans altered their surroundings not only to survive but also to prosper, early on making tools from stone, lighting fires, and hunting big game, then later planting and farming, and building cities and civilizations. Our rise has been swift.

Now, at six and a half billion strong and growing, humans possess more heft than any large animal in the history of the planet. On this last human species, the health of the earth now depends.

Because humanity itself has become a decisive force of nature, Nobel laureate Paul J. Crutzen describes our chapter in earth's biography as the "Anthropocene." He suggests that human-dominated time began in the eighteenth century, when James Watt invented the steam engine and coal-fired engines began warming the atmosphere with carbon dioxide emissions. As evidence of the Anthropocene, Crutzen points to human consumption of half earth's available freshwater; the damming of rivers (more water is held by dams than flows to the sea, spinning the earth faster, shortening the day); the destruction of rainforests and subsequent extinction of species; the use of fertilizer, which has doubled the planet's supply of fixed nitrogen; the release of toxic or severely damaging substances into the environment; and the rise of agriculture, cattle raising, and fossil fuel burning, which are warming the planet and increasing the concentration of greenhouse gases to the highest levels in the last 650,000 years. To this might be added that direct human alteration of the earth now covers 83 percent of the continents, and that our garbage, which began in Africa as a heap of cracked and discarded antelope bones—bones that most often were gnawed by rodents and decayed by microbes—has grown to include billions of pounds of plastic that is virtually indestructible.

For biologist E. O. Wilson, humanity's mark on the planet began not with the Industrial Revolution but long before, when *Homo sapiens* settled the continents and hunted large animals to extinction—giant tortoises and elephant birds, whose eggs were as large as soccer balls, on Madagascar; giant flightless birds and lizards in Australia; woolly mammoths in North America. It continued as agriculture evolved and farming replaced an abundance of plants with maize, rice, and wheat, and it continues today, as vanishing habitats, introduced species, pollution, continued culling of large animals, and global warming threaten 25 percent of earth's species with extinction in the next fifty years, and 50 percent by 2100, inaugurating a planetary mass extinction unseen since the demise of the dinosaurs.

Now humanity has extended its reach to the sea. For more than three and a half billion years, the ocean has been earth's lifeline. Life was spawned in the sea. The ocean sheltered earth's first plants and animals, and created an environment where walking fish could come ashore. Climate changes wrought by the sea created evolutionary pathways for mammals and gave rise to our human ancestors. The sea still sustains us. Rushing currents supply water to the atmosphere's protective greenhouse and send rain to dry land, making our lives possible. Ever since the first rain fell to fill the first sea, and the first ocean basin opened and closed, the sea has been essential to life on earth.

The living partnership between earth and the life it nourishes has been rebalanced, as a single terrestrial species—whose time has been but a moment—has become powerful enough to modify the sea itself.

Even as we discover how vast the ocean, how complex and wondrous its workings, how rich and diverse its inhabitants—with discovery upon discovery unpredicted and unimagined—we are redesigning it. For so long we have believed the sea's resources to be unlimited, erroneously equating immensity with immutability. From our home on dry land, we are powerful enough to restructure marine food webs, powerful enough to alter the chemistry and temperature of seawater. Today characterization of so much of the sea—the fertile waters of continental shelves, the icy reaches of the poles, the dazzling diversity of coral reefs, the porous edges of estuaries, and even the open ocean—is incomplete without considering the contribution of humans, and considering how, in the grand sweep of time, that contribution might matter to humanity, to our fellow inhabitants on earth, and to the sea itself. We hold earth's life-giving waters in our hands.

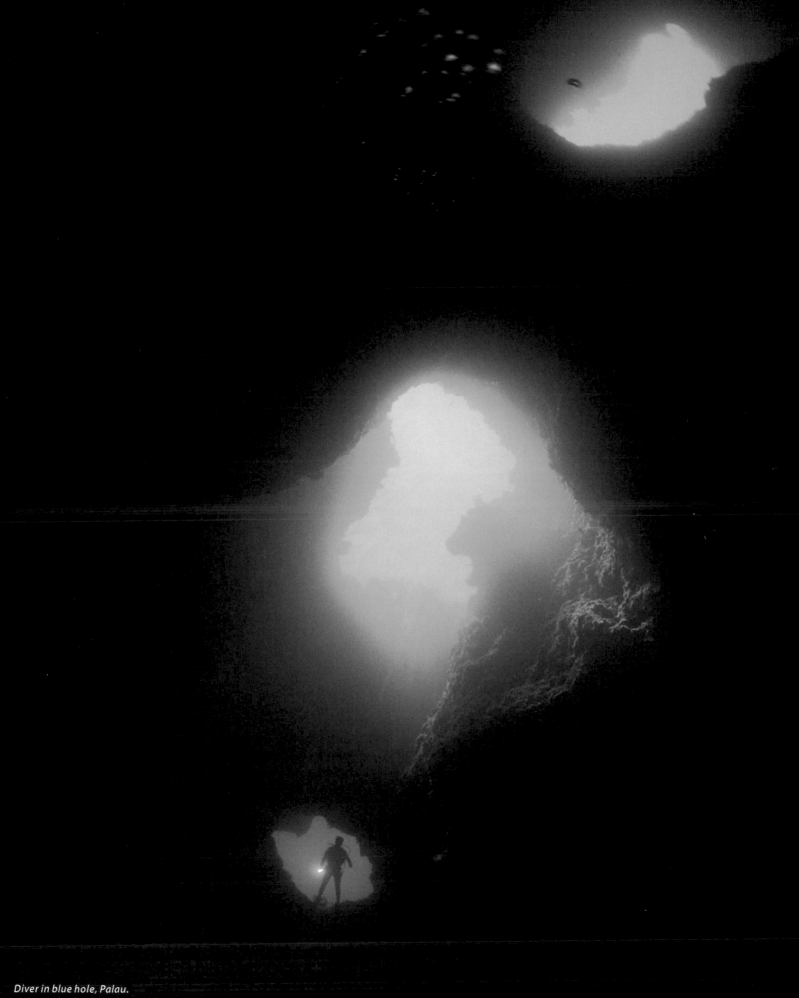

Diver in blue hole, Palau.
Only as we begin to understand the richness of the sea do we realize how much we are at risk of losing.

Coast at Rotsund, Norway. Human influence is now felt throughout the sea.

Touching the Sea

The Long Migration

6

In the winter, North Atlantic right whales come to calve in the warm waters of the Atlantic, off the Georgia sea islands. In the 1986–87 season, one whale, Stumpy, gave birth to a calf, Phoenix. They stayed in the calving grounds until spring, when mothers and calves swim north to rich feeding grounds in the Great South Channel and Cape Cod Bay. There the mothers gorge on thick concentrations of copepods, filtering them through their baleen. In the summer, the whales move on, many to the Bay of Fundy, to waters fertilized by surging tides. Each year pregnant right whales, fattened in their northern feeding grounds, return south to give birth. They are not the only marine animals to take such a long journey.

Every year thousands of whales and sea turtles make long migrations throughout the ocean. Gray whales ply the coast of the Pacific, traveling between summer feeding grounds along the ice edge in the Bering and Arctic's Chukchi Sea, and winter calving grounds in the warm lagoons of Baja California, Mexico. Humpbacks journey nearly 5,200 miles (8,300 km) between the coastal waters of the Antarctic Peninsula and Costa Rica, the longest migration known for any mammal. Pacific loggerheads hatched on the beaches of Japan cross the sea on the Kuroshio Current and spend their youth in the waters of the Baja, 6,200 miles (10,000 km) from the sands of their birth.

Whales and sea turtles travel the length and breadth of ocean basins. They have done so for millions of years, but now humans, relatively recent residents on earth, will determine whether their time is running out.

Below: Hippopo...
Kruger National...
Today hippopot...
closest living rel...

Opposite: Juven...
(Caretta caretta...
Palm Beach, Flo...
turtles make lo...
between foragi...
crossing entire o...
in their natal sa...

Above: Sperm whale (Physeter macrocephalus) pod, Azores, North Atlantic. Sperm whales, the largest of all toothed whales, are deep divers, seeking their prey of giant squid in waters 6,500 feet (2,000 m) deep.

All these whales had teeth. They swam in water that cooled as Australia drifted away from Antarctica, and Antarctica became covered in ice. Tiny krill flourished in the cool, nutrient-rich water. *Llanocetus*, whose skeleton was found on Seymour Island, Antarctica, had, in addition to teeth, the beginnings of baleen, fringed plates of hornlike material hanging from its jaw that strained krill from the water. *Llanocetus* is earth's oldest known filter-feeding whale.

Whales and turtles adapted to their water world, navigating it in ways we have yet to fully understand. Within hours of their birth, baby turtles leave their island beaches and return to the sea. Faithful to their natal home, they return as many as thirty years later to nest in the same sands. They make their way through seemingly uncharted seas, on a path once traveled but never forgotten, steering by internal compasses set long ago to coordinates on earth's magnetic field. Sperm whales, guided by nature's most powerful sonar system (housed in heads one-third the size of their bodies), home in on their prey of giant squid, in depths not easily accessible to their competitors. Male humpback whales sing to each other, eerie melodies with complicated phrasing, lasting as long as thirty minutes and changing with the seasons. The meaning of the songs remains elusive to us. The only other animal whose language shares a similar form of syntax is ours. When brain size is considered relative to brawn, the only brain larger than a sperm whale's is that of a human.

Above: Each year, thousands of green sea turtles (Chelonia mydas) return to breed on Raine Island, Great Barrier Reef, Queensland, Australia, one of their largest nesting sites.

Left: Humpback whale (Megaptera novaeangliae) singing, Maui, Hawaii. The meaning of the eerie songs of humpback whales remains elusive to us. (Photo obtained under NMFS permit #987)

Thick with Whales

Sailors on the European voyages of discovery describe a sea crowded not only with turtles, but also with whales. Jacques Cartier in 1535, in the Gulf of St. Lawrence, and George Best, at Baffin Island with Martin Frobisher in 1578, both observed whales as numerous as porpoises, which in those days meant groups of hundreds or thousands, many close to shore. The Basques ran what was at the time one of the world's largest whaling stations in Red Bay, Labrador. On Saddle Island, fragments of red roofing tiles, stones blackened with burned whale fat, and charred whale bones mark the site of the tryworks, where workers boiled down blubber from thousands of bowhead whales in copper cauldrons and poured off the oil into oak barrels.

In the nineteenth century, whale baleen was made into corset stays and umbrella ribs, and the oil was used in soap and lighting. Dr. Abraham Gesner, a Canadian physician and naturalist, saved sperm, right, and bowhead whales from outright extinction when he made kerosene and launched the petroleum industry, obviating further need for whale oil. Industrial whalers of the twentieth century killed whales for meat.

One thousand years of commercial whaling took a heavy toll on what is now the world's largest animal. Logbooks and capture data may underestimate the magnitude of the loss. DNA analysis, shining a light into that dark history, suggests that conventional estimates of whales before whaling are, for some whales, only 10 percent of their historical populations. Conventional estimates suggest that, before the onset of commercial

whaling, the sea was home to 115,000 humpbacks. Genetic analysis suggests the number may have been more like 1.5 million. Unable to see a world that disappeared before we could fully take its measure, we may, in setting goals for the restoration of endangered whales, be setting our sights too low.

PASSAGE OF PLASTIC

Plastic is durable. Designed to last, it persists long after it is thrown away, breaking down into smaller and smaller fragments. Humans manufacture millions of tons of plastic each year. When it is discarded, it doesn't disappear. Carried by currents and pushed by the wind, it travels far from land and accumulates in the sea. Fishing nets and gear, plastic bags, and packing material kill millions of seabirds, marine mammals, and sea turtles each year. Fur seals and sea turtles are entangled and strangled in nets, lines, and plastic collars. Sea turtles eat plastic bags, mistaking them for prey. Necropsy of a stranded sperm whale showed its stomach obstructed by a plastic garbage can liner, plastic sheeting, and a bread wrapper. Albatross and storm petrels ingest plastic granules and pellets. Some birds die, their intestinal tracts blocked. Others, like the red phalarope, grow thin: their stomachs filled with plastic, they have difficulty putting on the fat necessary for their long migrations.

Plastic has been found in the stomachs of fish, and even in the tissues of tiny filter-feeding salps. In a swirling mass of garbage located in the North Pacific, scientists found 6 pounds (2.7 kg) of plastic for every pound (.5 kg) of plankton. Plastic pellets trap highly toxic pesticides and industrial chemicals that then work their way through marine food webs, threatening the health of large marine animals and the people who consume them.

171

Left and right: Pacific humpback whales (Megaptera novaeangliae) off Alaska. Humpbacks mate and calve in warm tropical waters and migrate to cooler, more fertile higher-latitude waters to feed. Genetic analysis suggests that, before commercial whaling, their numbers may have been far greater than we had imagined.

Whaling Endangers More Than Whales

The loss of millions of whales reverberates through entire communities of marine animals, touching the foundation of marine food webs whose existence we have realized only recently. Those that dwell at the bottom of the sea depend for their sustenance on the rain of organic carbon falling from the surface. It is little; most of the sea's dead are eaten and recycled long before they reach bottom.

The sudden arrival of a whale carcass, a giant package of blubber, meat, and bones, is a feast for the hungry, an oasis of plenty in barren waters. A whale fall can equal up to two thousand years' worth of the slower, gentle rain of carbon from the sea surface. Scavengers, hagfish, and sleeper sharks feed on the carrion. A small whale provisions them for a few months or a year, a large blue whale for perhaps seven to eleven years. Bacteria feed on lipids trapped in the bones, producing hydrogen sulfide to nourish thousands of mussels. Clams, limpets, and snails move in.

In Clayoquot Sound off Vancouver, Canada, a gray whale (Eschrichtius robustus) *skims the sea surface.*

Thousands of worms pick apart the skeleton. One, the bone-sucking *Osedax*, takes oxygen from the water with its red feathery plume. Its fleshy green roots tunnel into the bone, breaking down the marrow to feed the worm, leaving the bones full of holes. The community of more than 200 hundred species fed and sheltered by the skeleton can endure for forty to eighty years. Some of its members can live at deep-sea hot springs and cold seeps; others live nowhere else.

A gray whale (Eschrichtius robustus) *swims through a canopy of kelp off the coast of California.*

Whaling may have emptied the deep sea of a rich source of nourishment, impoverishing vast stretches of seafloor. Between 1920 and 1980, an estimated two million great whales were taken from the sea. Entire seafloor communities are likely to have disappeared with them—communities that were diverse and vibrant in and of themselves, and that also may have served as stepping-stones for larvae of vent animals riding on deep currents. Of the species that live only at whale falls, between 20 and 50 percent may be on their way to extinction.

Many are those whose lives may have depended on whales. Untold numbers of California condors living along the coast may have fed on the carrion of stranded whales. Whaling, by depleting this source of food, may have accelerated the decline of this now critically endangered bird. Gray whales suction the seafloor as they feed, creating a slick of nutrients and tiny crustaceans at the surface, where seabirds come to feed. The population of Pacific gray whales before whaling—potentially three to five times higher than it is today—could have helped feed more than a million fulmars, phalaropes, kittiwakes, and murres.

Back from the Edge of Extinction

Kemp's ridley sea turtle (Lepidochelys kempii) hatchling swims near its nesting beach at Rancho Nuevo, Mexico. Kemp's ridley hatchlings live in the currents of the Gulf of Mexico and the Atlantic for several years before returning to coastal waters. Hatchlings mature when they are ten to seventeen years old.

Down to only a few hundred reproducing females, the outlook for the Kemp's ridley sea turtle had become grim by the mid-1970s. To save the turtle, Rancho Nuevo was designated a sanctuary, where eggs and nests were protected from poachers. U.S. shrimpers were banned from Mexican waters during the nesting season and required to use turtle excluder devices—gear that would release trapped turtles. The sale of turtle products was prohibited. Eggs were incubated on a Texas beach once frequented by Kemp's ridleys, in the hope that baby turtles would "imprint" on that beach and return there to nest when they reached maturity.

Though Archie Carr never lived to see realized a possibility he had imagined, others have. More than forty years of dedication and perseverance have begun to save the Kemp's ridley. In 2007, 5,500 female turtles came to nest on the beaches of Tamaulipas, and thirty to Padre Island. Though the Kemp's ridley's future is not yet assured, today biologists see the very real possibility in the not so distant future of a 10,000-turtle *arribada* on the beaches in Mexico.

Whither the Right Whale?

Once they were protected from commercial whaling, populations of breaching, water-slapping, playful humpbacks began to recover, and their numbers are still rising. Southern right whales summer in the fertile waters of Antarctica and calve off the coast of South Africa. The whales are healthy, and their numbers, though still small—about eight thousand—are also steadily growing.

For other right whales, the future is uncertain. The population of right whales in the North Pacific is only a fraction of its former size; off the west coast of North America, only a few remain. In the North Atlantic, right whales calving in Cintra Bay off the coast of Africa and migrating through the Bay of Biscay were among the first to be targeted by Basque whalers. Now only a few dozen remain; in all likelihood, that population will become extinct in our lifetimes.

A North Atlantic right whale (Eubalaena glacialis) *in Canada's Bay of Fundy. North Atlantic right whales live in an essentially urban ocean, their future poised between life and death.*

The last hope for the North Atlantic right whale rests with Phoenix and the other 400 right whales of the western North Atlantic. A whaling ban cannot by itself bring this whale back from the edge of extinction; whaling is only one way to take whales. Today right whales live in an essentially urban ocean, facing life-threatening dangers created by humans that live onshore. The sea is noisy. The voices of whales, drowned by engine noise from ships and fishing boats, no longer carry as far as they once did, making it harder to find mates and food. The sea is congested. Hundreds of tankers, containerships, and car carriers regularly pass through the calving grounds off Florida and Georgia and the feeding grounds in Massachusetts's Great South Channel. As whales migrate, they cross busy shipping lanes and swim through a sea crammed with millions of lobster traps and miles of gill nets.

In February 2004, Phoenix's mother, pregnant with her fifth calf, was struck by a ship off the coast of Virginia and killed. When Phoenix was ten years old, she was entangled and severely injured by fishing gear. Entangled whales thrash in the line, and if they can't shake it off, they drag it along. Hundreds of feet of line, crisscrossed around their flippers, looped around their bodies, or caught in their mouths, impede swimming and feeding and, if the line is embedded, expose them to infection. The percentage of whales caught in fishing gear is rising: today fishing gear entangles more than three-quarters of North Atlantic right whales at least once in their lives.

The North Atlantic right whale is poised between life and death; every premature death edges the species closer to extinction. The Endangered Species Act forbids humans to kill any right whales, even accidentally, but year after year they die. Between 1980 and 1995 the average life expectancy of a female right whale dropped from fifty-two years to fourteen. The coastal waters of the North Atlantic are unsafe for right whales, and, to date, measures taken to protect them have neither reduced the rate of entanglements nor the deaths from ship strikes. Phoenix's mother died, as did one of her children and her unborn calf, from the trauma of ship strikes. Phoenix, her sister, and her daughter Smoke have already been entangled in fishing gear. Between 2001 and 2006 at least twelve right whales died from fishing entanglements and ship strikes.

Below: The North Atlantic right whale (Eubalaena glacialis) *named Phoenix in the calving grounds off the coast of Georgia.*

Opposite: In the North Atlantic, migrating whales are at risk of being struck by ships and entangled in fishing gear. Top: North Atlantic right whale (Eubalaena glacialis) *in the Bay of Fundy, Canada. Middle and bottom: Humpback whales* (Megaptera novaeangliae) *tangled in a fishing line off Newfoundland, Canada.*

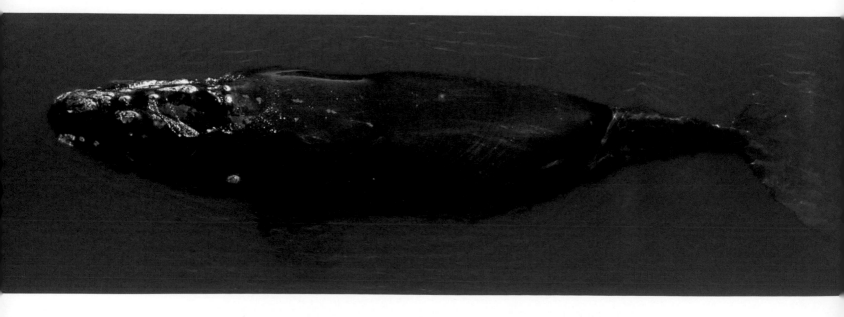

See Map 12, Shipping Lanes and Major Ports, p. 271; and Map 16, Lethal Whale Strikes, p. 273.

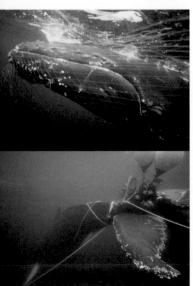

These deaths can be prevented. It was possible to save the Kemp's ridley, and it is possible to save the right whale. We can return to right whales their ocean home, make it safe once again, and renew their earthly lease. Saving two mothers each year from lethal ship strikes or entanglements can rebuild their dwindling numbers. Phoenix has a sister, two daughters, Smoke and Fuse, and a granddaughter alive and well. If they can journey unharmed between fertile feeding areas in the Gulf of Maine and the warm calving area off Georgia and Florida, together they could add at least twenty-six whales to the population and help carry the group back from extinction.

Phoenix, named after the mythic bird that was reborn from the ashes, represents hope for critically endangered right whales and the possibility of renewal for a beleaguered species. The bird, though, was solitary, rising from its burned nest unassisted. Phoenix and her family cannot endure without our help.

See Map 15, North Atlantic Right Whale Feeding and Calving Grounds, p. 273.

Edge of Continents

Humanity's romance with the sea began early in our history, when Homo sapiens, living in seaside cliffs overlooking the Indian Ocean in what is now South Africa, began consuming seafood.

Early in their history, humans were drawn to the sea. High in a cliff overlooking the Indian Ocean is a cave where they once dwelled, and in it, an ancient hearth. Scattered amid the ashes are tiny flaked stone blades, pieces of ocher, some cut and ground into powder to make red paint, and seashells. Today the cave, on South Africa's Pinnacle Point, lies at the water's edge. Back when the cave was inhabited, glaciers had cooled the climate and the sea had receded, exposing a shore 2.5 to 3 miles (4 to 5 km) wide, a good hike from the cave.

Food was scarce then. Fruit, tubers, roots, and antelope meat were not so plentiful in the drying, arid climate. Seeking sustenance, the cave dwellers turned to the sea, in one of the last expansions of the human diet before the cultivation of crops thousands of years later. The early humans foraged for shellfish amid the rocks and tide pools, consuming brown mussels, whelks, and giant periwinkles. A whale barnacle left in the cave suggests they may even have scraped blubber off a beached whale.

So began, 167,000 years ago, on a rocky shore now submerged beneath the water, humanity's romance with the sea. Early humans were sustained by seafood during the cold climate of the glaciers and perhaps along a long coastal migration from Asia into the Americas. Human exploitation of the sea, begun at Pinnacle Point, continues today. Extracting more than 84 million metric tons of seafood from the ocean each year, we are altering the sea in ways we have only begun to recognize.

Window into the Past

The fossil record of this recent time in earth's history, when humans take large quantities of fish and shellfish from the sea, is only now being written. Those who follow us will find it someday in the peaks of new mountains, in the walls of new quarries. For now, the history lies in the archaeological record of fish bones and midden heaps, in genetic analyses, and in the written record, a record whose distant origins lie in Pinnacle Cave.

The use of symbolic expression distinguishes humans from other animals. Some archaeologists believe it may have originated when greater food security lessened the demands of hunting and gathering. One of the earliest forms of symbolic expression—red paint ground from ocher in the cave at Pinnacle Point—may have arisen from freedom provided by an abundant and reliable supply of seafood in the Indian Ocean. Many thousands of years later, symbolic expression would evolve into writing, which, along with fish bones, shells, hooks, and other fishing gear from archaeological sites, has created a critical historical record. That record, of ship logs and manifests, catch reports, diaries of conquistadores and naturalists, railway inventories, accounts of bankers and financiers, tithes paid to the church and taxes to the state, is now being uncovered. It describes a fertile ocean brimming with fish.

Humans continue to extract fish and shellfish from the sea—for food, for fish meal, and for aquaculture. Below: Fishermen in the Ganges delta set nets to catch shrimp larvae coming in from the sea.

Sea of Plenty

The ocean's most fertile waters lie over the shelves of continents. In northern Europe and North America, they bear the mark of glaciers. The floors of the North Sea and the Gulf of Maine are hilly, strewn with sand, gravel, and big boulders left by receding ice. Rivers cut through the rubble, sculpting deep channels and high banks—Dogger Bank in the North Sea, Georges and Stellwagen banks and Jeffreys Ledge in the Gulf of Maine. They would become ideal homes for fish when the sea returned.

The North Atlantic's continental shelves, sculpted by glaciers, have nourished some of the world's most productive fisheries. Above: High surf at Schoodic Point in Frenchman's Bay, Maine, Atlantic Ocean.

As the ice melted, 11,500 years ago, the banks became forested islands where mastodon and moose roamed and people dwelled. By 6,000 years ago, the rising sea had submerged the land. Currents and tides churned the sunlit waters, mixing in nutrients that would fuel some of the world's most productive fisheries. On land edged by productive salt marshes, estuaries nourished small menhaden and herring that would feed big cod.

Alewives (Alosa pseudoharengus) at Merrymeeting Bay, Kennebec Estuary, Maine. Every spring, New England's estuaries were once clogged with millions of alewives swimming in from the sea to spawn in freshwater.

In eleventh-century Europe, as garbage, sewage, and silt from soil erosion polluted freshwater streams and rivers, and once plentiful salmon, whitefish, and eel runs were fished out, medieval fishermen set out to sea. They found unfathomable plenty. Fish were abundant—and big. A fish bone tossed on the floor of an English tavern and unearthed 900 years later belonged to a giant cod. Viking deposits in Scotland were filled with cod bones, many twice the size of those caught today.

Across the ocean, the quantities of cod were staggering. Reports from John Cabot's 1496 voyage to Newfoundland were of cod so plentiful they could be captured with baskets, and so bountiful they slowed his ships. One of the first governors in Newfoundland described a sea so thick with cod that it was almost impossible to row a boat through. A French priest out on Newfoundland's Grand Banks in 1719 found cod as numerous as the grains of sand on the seafloor below, and as valuable as South America's great gold and silver mines.

These early reports of astonishing plenty can seem the stuff of dreams, but fisheries scientists and historians piecing together the longer historical perspective are unearthing account after account of waters crowded with fish and shellfish: a Native American midden on Maine's Damariscotta River, larger than a football field and stacked 16 feet (5 m) high with huge oyster shells; in Saugus, Massachusetts, enough mackerel driven ashore by feeding stripers that men carted them away in wheelbarrows; Maine fishermen harpooning 300- to 400-pound (135- to 180-kg) swordfish just off the coast, some in Portland Harbor, within sight of land.

Along with millions of cod, mackerel, and oysters, New England's waters were home to whales, including the beluga, as well as large flightless great auks, and 20-pound (9-kg) lobsters that crawled in the rocks along the shore at low tide. Wherever rivers met the sea, on the Merrimac, the Piscataqua, and the Saco, large numbers of cod came to feed on herring and menhaden, alewives and shad. When fishermen blocked the rivers with weirs, they caught millions of pounds of alewives. There are records of dory fishermen just outside Gloucester Harbor catching 400 to 600 (180 to 270 kg) pounds of cod in the morning and returning home by mid-afternoon. Farther out, on Stellwagen Bank, they took 12,000 pounds (5,400 kg) of cod in a day. Gillnetters took thousands of pounds of cod near shore in Ipswich Bay.

Atlantic cod (Gadus morhua), Norway. Cod, once plentiful in the North Sea, is but a fraction of its former abundance.

Empty Ocean

This great abundance would not last. As rivers were dammed and fishing intensified, populations of alewives and cod dwindled. As early as the middle of the seventeenth century, Massachusetts fishermen failed to meet their contracts for cod deliveries to fish merchants. Laws appeared regulating the use of weirs and prohibiting commercial catches of spawning cod. The regulations were too little and too late. Today nearly half the cod spawning grounds in the Gulf of Maine have disappeared.

Nineteenth-century logbooks and catch records suggest the level of decimation. In 1852 forty-three schooners out of Beverly, Massachusetts, fished for cod along the Scotian Shelf south of Nova Scotia. They fished with hand lines and hooks. The entire fleet, using fewer than 1,200 hooks, caught more cod in 1852 than ninety modern vessels did from a much larger area (that included the Scotian Shelf) in 1999, even though the modern vessels were equipped with sonar to find fish and huge nets to catch them. A once great fishery had been depleted. By 2002 stocks of adult cod on the Scotian Shelf had collapsed to a mere 0.3 percent of what they had been in 1852.

As each fishery declined, fishermen moved into deeper water, to the steep edge of the continents and to seamounts—extinct volcanoes rising from the floor of the deep sea. These waters, only 0.0001 percent of which have been studied, are inhabited by slow-growing, long-lived fish: orange roughy, roundnose grenadiers, and, in Antarctic waters, the Patagonian toothfish. They too are quickly being fished out.

Disappearing Giants of the Sea

In the eastern Atlantic and Mediterranean, the number of bluefin tuna has plummeted. The population of these giants of the sea, majestic and powerful swimmers that cross entire ocean basins, is in imminent danger of collapse. Local populations are becoming extinct, and their geographic range is shrinking. Today it's rare to find a bluefin of any size in the North Sea, but it hasn't always been that way. In the early 1900s, thousands of bluefin surged into the North Sea each summer to feed. They chased garfish right into shore, to waters just outside a Danish castle. They fed on great schools of herring on Dogger Bank and in the Kattegat, a narrow arm of the North Sea between Denmark and Sweden. They weighed between 110 and 220 pounds (50 and 100 kg); some were 1,500-pound (700-kg) giants. In the 1920s, fishermen began targeting them, every year taking thousands of tons of bluefin from the North Sea. By 1970 the fish were gone. Fishermen hunted them in other waters, and today the ocean is almost empty of bluefin.

Great sharks are also disappearing. Sharks are skilled hunters that use seven senses: the five we have, plus additional sensitivities to pressure and electrical currents that help detect the distant motions of prey. Sharks grow slowly, mature late, and, unlike other fish, give birth to live young. Dusky sharks mature when they are twenty years old, and sand tiger sharks birth two live pups at a time. Sharks migrate across entire ocean basins; coastal lagoons and estuaries are their nurseries. Though they are the sea's apex predators, we are taking many of them faster than they can reproduce.

See Map 13, Fish Stock Exploitation, p. 271; and Map 14, Change in Fish Biomass, p. 272.

Top: Yellowfin tuna (Thunnus albacares) hooked during sport fishing in the Pacific off the coast of San Diego, California. Yellowfin live in warm water throughout the world ocean. Yellowfin caught by trolling or with pole-and-line gear yield lower bycatch—incidental catch of sea turtles, seabirds, and dolphins—than purse seines or longlines.

Above: Sharks are deliberately targeted for their fins. Here multiple species of shark fins are hung to dry at a shark finning camp in Baja, Mexico, Sea of Cortez.

Millions of sharks are killed each year, accidentally hooked on longlines intended for tuna or swordfish or deliberately targeted for their meat, vitamin-rich oil, and highly profitable fins, which can sell for over $300 per pound ($700 per kg) in the shark fin soup market. We regularly identify new species of shark: the 18-foot (5.5-m) -long megamouth, which, despite its large size, feeds primarily on tiny krill, and which has been seen only forty times, most often in fishing nets; six new species of bioluminescent lantern sharks; and a new hammerhead. Many others may disappear before we identify them.

Believing the sea's resources to be infinite, its abundance inexhaustible, we are emptying the ocean. The loss is catastrophic. Scientists estimate that overall, in the last fifty to hundred years, fishing has depleted large communities of fish—cod, swordfish, halibut, tuna, and sharks—by 90 percent. For some the devastation is great: large sharks, skates, rays, and cod in the North Sea; oceanic whitetip sharks in the Gulf of Mexico; onion-eye and roundnose grenadiers in the North Atlantic (fish that live for about sixty years and mature in their late teens, and that in a very short time may become extinct); sandbar, blacktip, tiger, bull, and scalloped hammerhead sharks off the coast of North Carolina. Critically endangered sawfish, whose noses resemble chainsaws, have already disappeared from many coastal waters. Sharks have lived in the sea for 400 million years. Now they may not survive us.

Tiger shark (Galeocerdo cuvier), Bimini, Bahamas. Living for as many as fifty years, growing longer than 18 feet (5.5 m), tiger sharks are valued for their flesh, fins, skin, and liver oil.

Below left: The aquarium fishery has threatened the Banggai cardinalfish (Pterapogon kauderni) from Indonesia with extinction. Below right: Hawaiian monk seal (Monachus schauinslandi), yearling female, Mahukona, Kohala, Hawaii. Only 1,200 Hawaiian monk seals remain. Starvation, entanglement in marine debris, predation, disease, and erosion of pupping beaches threaten this species with extinction.

Opposite, top left: Great hammerhead shark (Sphyrna mokarran), Northern Bahamas. Named for their hammer-shaped heads, these endangered sharks, averaging over 500 pounds (230 kg) and reproducing only once every two years, are valued for their fins. Top right: Endangered West Indian manatee in Florida's Crystal River. Bottom right: Goliath grouper (Epinephelus itajara) surrounded by a cloud of round scad (Decapterus punctatus) during fall spawning off the coast of Jupiter, Florida. Critically endangered but now protected, the goliath grouper is showing promising signs of recovery. Below left: Manta rays (Manta birostris) feeding along a plankton line in waters off the Marquesas Islands in French Polynesia. Unfished populations of manta rays are not threatened, but their numbers are declining where these large, slow-moving fish are taken for medicinal purposes and for their fins, livers, and meat.

We humans measure our time in hours, days, months, and years—a span so short it has been difficult to grasp that we may be initiating earth's sixth mass extinction. Giant marsupials and giant monitor lizards (twice the size of Komodo dragons) vanished following our earlier settlement in Australia. Mammoths and mastodons, giant bears, saber-toothed cats, and elephants once roamed America; their disappearance coincided with our arrival thirteen thousand years ago. We have begun to record extinction in the sea resulting from our actions. In the eighteenth century, the Bering Sea was filled with herds of Steller's sea cows. European hunters extirpated them in twenty-seven years. One summer day in 1844, Icelandic hunters killed the last great auks, large flightless seabirds that once numbered in the millions. When Columbus arrived in the Caribbean in 1492, he hunted the Caribbean monk seal, the sea's only native seal. It was last seen in 1952. In 1973 an extensive aerial survey of islands and atolls in the Gulf of Mexico and the Caribbean found no evidence of seals. After another extensive search in 1984 for seal scat, tracks, carcasses, or bones came up empty-handed, and lighthouse keepers reported no sightings, the Caribbean monk seal was declared extinct.

The passage toward extinction begins when, year after year, populations become depleted, as so many animals at the top of the sea's marine food webs are today. It continues as populations become so impoverished they cannot easily replenish themselves. Eventually they become regionally extinct, like the Atlantic gray whale, or ecologically extinct, like sharks whose numbers have dropped so low they no longer fully fill their roles as apex predators. Extinction can happen in a flash, as it did for the Steller's sea cow, or it may take several hundred years, as may happen with the North Atlantic right whale.

Earth's previous mass extinctions were catastrophic in the number of species, genera, and families snuffed out, but the timing may have been leisurely. The Great Dying of the Permian extinction, when 90 percent of marine life disappeared, took place over 160,000 years. The asteroid that hit earth at the end of the Cretaceous was the coup de grâce for an extinction that had been going on for 100,000 years.

In our time, we humans, now earth's most dangerous predator, are accelerating the pace of extinction. Already a hundred times higher than before our arrival, it will become, if current trends continue, a thousand times greater in the next decades. The archaeological and historical record indicates that we are depleting our ocean. Uniquely able to look into our past and observe our present, we, the last human species, now face the question: what is our responsibility to earth's creation, to the sea that sustains us?

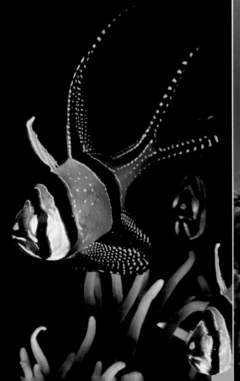

Cascading Losses

The loss of the sea's top predators cascades throughout marine food webs. For four thousand years in the Gulf of Maine, cod, haddock, halibut, and pollock thrived, keeping sea urchin, snow crab, and shrimp populations in check and allowing kelp forests to flourish. When predatory fish were taken, populations of prey surged, and fishermen began fishing down the food web, profitably taking snow crab and shrimp. Sea urchins decimated the kelp until they, too, were extracted, and the kelp returned in a rearranged and diminished ecosystem.

For thousands of years, Steller's sea cows, sea urchins, and sea otters lived in the kelp beds of the Northwest Pacific. Sea otters preyed on urchins, preventing them from grazing away kelp, until fur traders hunted otters to the edge of extinction. Sea urchins then denuded the kelp, until otters were protected and urchins were fished. There are few marine food webs where our signature cannot be found.

Trawlers drag their gear over the seafloor, uprooting slow-growing communities of coral, mussels, and anemones living there, scouring the seabed, leaving it bare. When the gear sweeps through frequently, the longer-lived organisms can't recover, and the communities they support cannot be reestablished.

Below: Pacific cownose rays (Rhinoptera steindachneri) give birth to one pup per year, leaving the fish susceptible to overexploitation.

Opposite top: California sea otter (Enhydra lutris) in a Pacific giant kelp forest off the coast of Monterey, California. Human competition for their shellfish prey, entanglement in fishing gear, oil spills, and predation by killer whales threaten California sea otters.

Opposite middle: Fishermen from the village of Brenu-Akyinimt, Ghana, head out to sea. Opposite bottom: Fish caught with nets from dugout canoes and hauled in by hand in Sulima, Sierra Leone. Fish stocks off the coast of Northwest Africa, exploited by foreign fishing fleets, have plummeted.

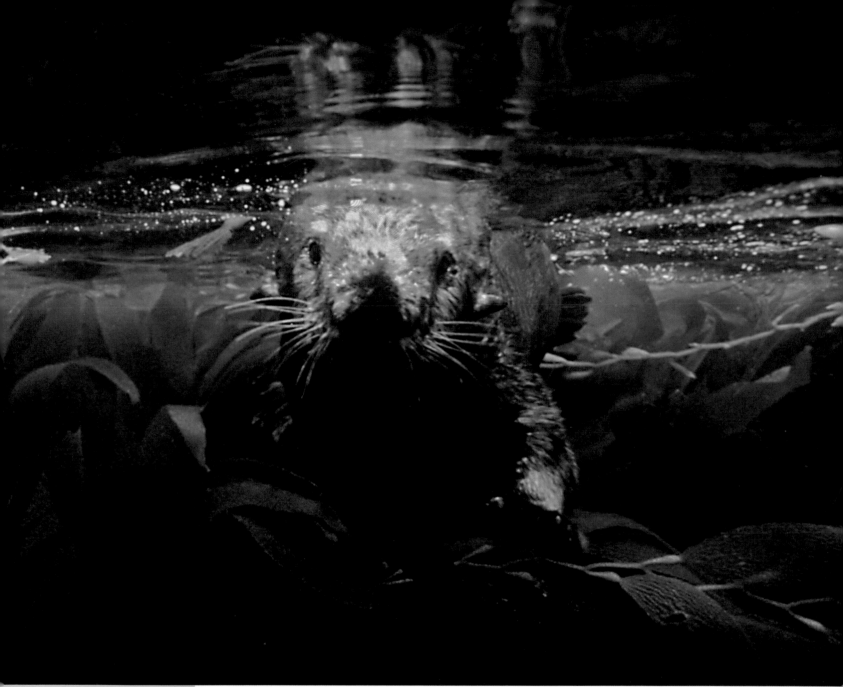

Near the bottom of marine food webs are jellyfish that proliferate when large fish are removed. The fertile, nutrient-rich waters of the Benguela Current, off the coast of Namibia, once supported vast numbers of anchovies and sardines. Fishermen removed them, the stocks collapsed, and 12.2 million metric tons of large jellyfish, whose umbrellas span between 5 and 10 inches (13 and 26 cm) replaced them.

Replacing fish with jellyfish threatens the food security of 2.6 billion people who depend on fish for protein. In Ghana the bushmeat trade is directly linked to the fertility of the sea. When fish catches are high, demand for bushmeat falls. When the catch is poor, demand for bushmeat rises. Foreign exploitation has substantially reduced fish stocks in the Gulf of Guinea. Mammal populations in Ghana's nature reserves have declined by 76 percent, with some species becoming locally extinct. The sea is an essential source of food, and its resources are renewable. Managed well, it can be a reliable source of healthy food.

Farming the Sea

Workers tend an oyster/pearl farm off the coast of Western Australia.

Over time, as hunting and foraging no longer met the growing demand for food, humans domesticated land plants and animals, turning wild forests and prairie into cultivated fields and industrial animal farms. Now, as the population continues to grow, and collapsed and depleted wild fisheries no longer meet increasing demand for fish, humans domesticate the sea. Aquaculture now supplies over 28 percent of the world catch (by weight) of marine fish and shellfish.

Rapidly growing and inadequately regulated, fish farming can endanger wild fish. Farmers in the Bay of Bengal, netting young tiger shrimp to stock aquaculture ponds, throw away millions of young wild fish and shrimp each year. Deadly viruses from shrimp farms spread to wild populations. In the inlets of British Columbia near Vancouver Island, outbreaks of lice in salmon farms are killing wild salmon. Within four generations, wild salmon in these rivers will disappear. Massive amounts of wild fish go into fish meal and fish oil to feed carnivorous shrimp and salmon. Between 4 and 7 pounds (2 to 3 kg) of wild fish such as herring, sardine, and mackerel are required to produce 2 pounds (1 kg) of farmed salmon or shrimp. In the North Sea, extraction of capelin and sand eel for fish farms threatens the food supply of cod. The United Nations Millennium Ecosystem Assessment finds that the cost of aquaculture to wild fisheries is high.

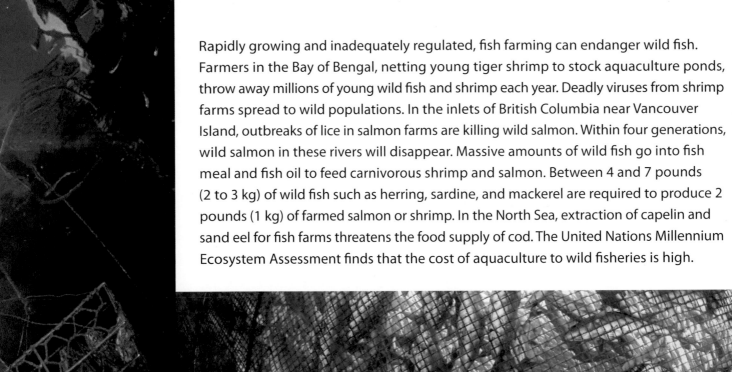

Top right: Almaco jack (Seriola rivoliana) in a pen at a fish farm off the Big Island, Hawaii.

Bottom right: Salmon breeding pens, Lofoten Islands, Norway, Atlantic.

A Future for Wild Fish

The historic record now describes in rich detail the sea's past abundance and its present vulnerability. In the Northwest Atlantic, off the coast of Canada, cod that supported a prolific and lucrative fishery for almost 500 years were depleted by as much as 99.9 percent. The stock collapsed, putting thousands of people out of work and draining the Newfoundland economy of millions of dollars. It has yet to recover.

If present trends continue, the ocean's once vast fisheries may collapse by 2048. This needn't happen. The yield of historical research is that the sea's capacity is much higher than we ever realized. The promise of contemporary fisheries science is the possibility of returning the sea's fish to health and abundance. The tools and knowledge are there. Too many fishermen still chase too few fish: subsidies have created a global fishing fleet 250 percent larger than what is needed. Big mothers matter: older fish produce exponentially more eggs than younger fish, but they rarely live to spawn more than once. Shark and tuna make entire ocean basins their home, oblivious to boundaries between nations. Their health depends on worldwide fishing restrictions and trade sanctions that prevent their sale for soup and sushi. Fisheries scientists know how to protect fish; for too long, politicians and fishery managers have ignored or diluted their recommendations.

We will determine whether our children and grandchildren will know a sea of wild fish. Opposite: Whale shark (Rhincodon typus), earth's largest shark, on Ningaloo Reef, Western Australia. Below left: Oceanic whitetip shark (Carcharhinus longimanus), Bahamas. Below right: Male Atlantic salmon (Salmo salar) fertilizing eggs in the St. Jean River, Québec, Canada.

When fishing is curtailed before a population completely collapses, fish can recover. Overfished seas refilled twice in the respite forced by two world wars. When a lawsuit forced the closure of parts of Georges Bank in 1994, scallop beds were rejuvenated, growing, within four years, to fourteen times their density before the closure. Haddock experienced an exceptional year on Georges Bank in 2003. Record numbers were born, and enough survived so that if they are allowed to spawn repeatedly, they can rebuild the population. On the Pacific Coast, strict management has returned once-depleted Alaskan wild salmon fisheries to health. Perhaps it is time, before the rest of the sea is emptied, to let it rest and renew itself. Our own health and prosperity depend upon it.

8 Rhythms on a Reef

Gorgonian sea fan and anthias fish on remote reefs of the Eastern Fields, in the Pacific's Coral Sea.

A coral reef hums with energy. Jet black damselfish farm their own algal garden patches, zealously guarding them against invaders and weeding out unpalatable algae. Powerful mantis shrimp take deadly aim at their prey, impaling fish or smashing mollusk shells. Resplendent lionfish with bold stripes and venomous spines peacefully allow brightly colored wrasses to pick off their parasites. Yellow- and purple-striped clownfish find refuge in the stinging tentacles of anemones. Parrotfish, groupers, and wrasses change sex as population size and need dictate. A nurse shark sucks fish out from under the sand.

Even the stone breathes. Amid the bustle on the reef, corals are quietly building what may be earth's largest living structures. The enterprise is based on a symbiotic relationship with tiny, single-celled algae, zooxanthellae, living within the tissues of their coral hosts. In return for nutrients and shelter, the photosynthesizing algae turn carbon dioxide and light into carbohydrates, satisfying corals' energy needs and enabling them to build massive limestone reefs. Corals thrive in clear, warm, impoverished water, where they and their algae feed each other in an endless cycle.

On spring nights on Australia's Great Barrier Reef and in late summer on Caribbean reefs, just after the full moon, corals spawn, filling the sea with swirling clouds of eggs and sperm. Coral reefs are the most diverse and biologically complex ecosystems in the ocean, and for years, the exquisite timing and trigger of broadcast spawning remained a mystery. Day by day, year by year, our understanding of the seemingly infinite variety and complex interactions among reef inhabitants expands. At the same time the reefs themselves are contracting.

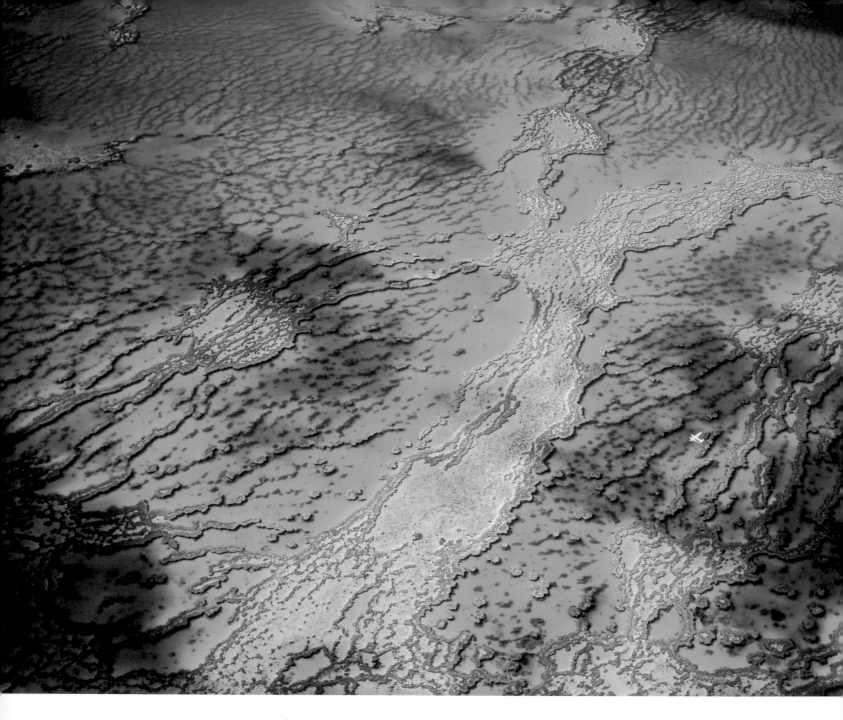

A Reef to Rival a Rainforest

Above: Aerial view of the Pacific Ocean's Great Barrier Reef Marine Park, Queensland, Australia, home to hundreds of species of fish, mollusks, and coral, including, far right, schooling hussar (Lutjanus adetii).

Opposite left: Lava river, Hawaii Volcanoes National Park, Big Island, Hawaii. Live coral covers more than half the waters surrounding Hawaii's volcanic Big Island.

The thousands of islands and reefs of Australia's Great Barrier Reef are separated from the mainland by a blue lagoon. Other reefs fringe the shore. Reefs of the Hawaiian Islands, far from the edge of continents, grow around volcanoes. Some still belch black smoke and lava. As the volcanoes cool and subside, the coral ring remains, surrounding a lagoon with the volcanic peak at its center. Eventually the volcano sinks below the water line, the only evidence of its former presence a coral atoll. In the vast ocean, coral reefs constitute only 0.2 percent of the seafloor. That small space is a hotspot of marine diversity, home to one-quarter of all known marine life, most of it still to be named. Second only to rainforests, coral reefs may host between one and nine million species of marine organisms—including hundreds of species of coral and thousands of species of fish.

See Map 6, Coral Reefs, p. 268.

Reef-dwelling cardinalfish (Apogon fleurieu) from the Red Sea brood their eggs in their cheeks.

Below: Feather star crinoid (Comanthina sp.) on gorgonian sea fan (Subergorgia mollis) in Papua New Guinea's Kimbe Bay, home to some of earth's most biologically diverse coral reefs. Bottom: A cowry (Cypraea sp.) blends in with the pink-and-white soft coral on which it grazes, Raja Ampat Islands, Irian Jaya, Indonesia.

The mysteries of reefs continue to unfold. Just recently, scientists learned that corals can "see." Special light sensors, tuned to the dim light of the moon, cue corals on a reef to spawn in almost perfect synchrony, on the same night at roughly the same time, within a few hours. The full riches of coral reefs have yet to be measured. Scientists surveying the submerged islands and reefs of the Bird's Head Seascape, off the coast of Papua, Indonesia, found a shark that walks on its fins. Scientists diving in the French Frigate Shoals in the Northwestern Hawaiian Islands are finding coral, crabs, sea squirts, and sea stars that may, when they are fully analyzed, prove new to science.

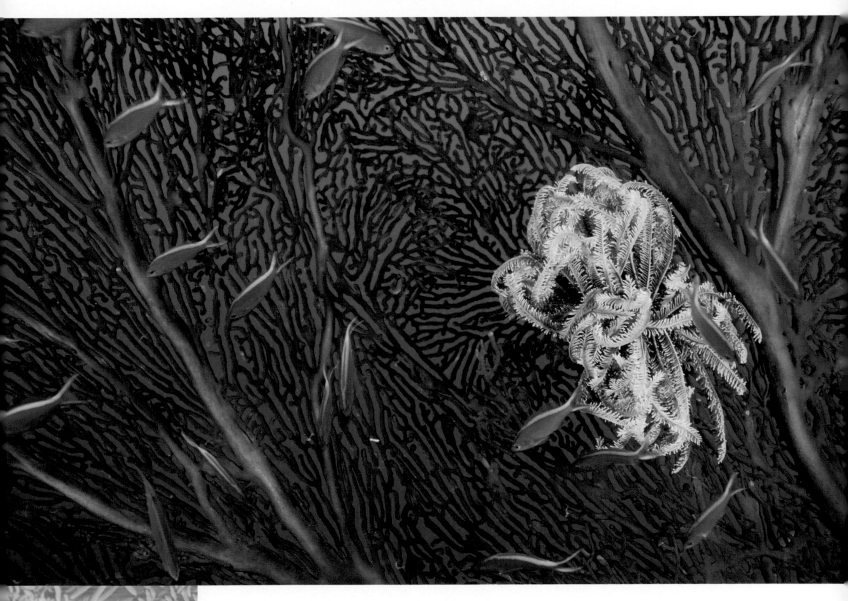

Below: Feather star crinoid (Comanthina sp.) on gorgonian sea fan (Subergorgia mollis) in Papua New Guinea's Kimbe Bay, home to some of earth's most biologically diverse coral reefs. Bottom: A cowry (Cypraea sp.) blends in with the pink-and-white soft coral on which it grazes, Raja Ampat Islands, Irian Jaya, Indonesia.

Among these diverse organisms may come the next generations of life-saving medicines. Researchers have barely begun to inventory the chemical arsenal that reef dwellers employ to fend off predators, catch prey, or defend their turf, but already anticancer drugs have been found in reef sponges and anti-inflammatory agents have been discovered in gorgonian corals. Coral reefs generate $30 billion each year for their fisheries, coastal protection, biodiversity, and tourism. Yet despite their intrinsic value as a repository for a large share of earth's diverse life, and despite the income they provide and the jobs they generate, they are disappearing. It would not be the first time.

Reefs through Time

Crinoids have inhabited earth's reefs for millions of years. These modern crinoids, the noble feather stars (Comanthina nobilis and Oxycomanthus bennetti) live amid soft coral (Dendronephthya sp.) in waters off Madang, Papua New Guinea.

Throughout earth's history, reefs have come and gone. Time and time again, they would grow, flourish, and then become extinct. The unfamiliar names of animals that built them—archaeocyathids, sea lilies, brachiopods, rudists—speak of a long line of reef builders that are no more. The first reefs, stromatolites, appeared some 3.5 billion years ago. They were built by mats of microbes that trapped sand and sediment from ocean currents. Grit built up, and the mat thickened. As the microbes moved up toward the light, their abandoned quarters were cemented into layers of rock preserved today in remote areas of western Australia.

Archaeocyathids, sponges that looked like nestled ice cream cones, began building reefs about 530 million years ago. They appeared in warm shallow water, where they multiplied and diversified, and built structures that would shelter some of the sea's first large animals. Between 10 and 15 million years later, they disappeared, their remains left throughout the world, including along Siberia's Aldan River and in the foothills of Nevada. Though their builders became extinct, reefs would reappear 25 million years later, constructed by new communities of animals.

Coral built earth's next great reefs, but not the coral we know today. Honeycombed tabulate coral, spongelike stromatoporoids—some nearly 10 feet (3 m) in diameter—and colonies of filter-feeding bryozoans built earth's reefs between 470 and 360 million years ago. Meadows of sea lilies grew around the reefs, while nautiloids (relatives of today's pearly nautilus) swam among the coral. These reefs, now beached in Alberta, Canada; Windjana Gorge in western Australia; Morocco; Algeria; Greenland; and along the Great Lakes in the United States, were more widespread than the reefs of today.

Earth's reefs flourished, declined, and then returned, weathering a Devonian mass extinction to build new reefs built of new communities of sponges, bryozoans, bivalved brachiopods, and rugose corals. The animals succumbed to the end-Permian extinction 250 million years ago, and the reefs became the Guadalupe Mountains of Texas.

The record of earth's early reef builders lies ashore, on seafloor that is now part of the land. The inhabitants of today's reefs, too, will one day leave their imprint on dry land. Left: A sea-horse (Hippocampus kuda) *on branching lace coral* (Sylaster sp.) *and a sea fan* (Melithaea sp.) *in waters off Manado, Indonesia. Right: Spawning barrel sponge* (Xestospongia testudinaria) *off Papua New Guinea.*

Following the Great Dying, several million years elapsed before life was restored to the sea. When reefs again reappeared, they were characterized by yet another group of organisms: new corals whose descendants would dominate today's reefs, and rudist clams, whose tenure would be shorter. In some rudists, their valves, one large and conical, the other small and flat, resembled garbage cans with lids. Some grew upright, their bottom valve anchored to the seafloor, the top poking above the sediment. As the sediment accumulated, they grew to keep pace. Others reclined, their heavy, horn-shaped shells steadying them in shifting sand and strong currents. Some grew to be more than 3 feet (1 m) long.

When a hurricane toppled the reef, they would rebuild upon the rubble. Rudists did not survive the mass extinction that killed the dinosaurs at the end of the Cretaceous, 65 million years ago, but the reef debris endured and became a reservoir for the spectacular oil fields of Oman and the Gulf of Mexico.

Dwellers of the sea make their homes in places that continually surprise us, repeatedly overturning our assumptions of how and where the sea supports life. It was long thought that corals inhabited warm, shallow waters of the tropics and that kelp forests grew in cold, temperate waters. Tropical kelp was considered a rare relic from a distant time when the water was cooler. One species, *Eisenia galapagensis*, had even been placed on the World Conservation Union (IUCN) endangered species list. Kelp can live in the clear water of the tropics, in places where cold and fertile water upwells within the deepest reach of the sun. In a deepwater refuge off Isla Fernandina and Isla Isabela in the Galápagos, scientists found an extensive kelp forest and, in it, grazing fish, sea hares, and decorator crabs formerly thought to live farther north in temperate waters.

Cool Coral

Above: The kelp Eisenia galapagensis *lives in a deepwater refuge off the Galápagos.*

Below: Coral grows in cold ocean water. Left: A sea fan (Paramuricea placomus) and amphipod (Epimera cornigera) in Trondheims Fjord, Norway. Right: Soft coral (Capnella glomeratum) from Svalbard, North Atlantic.

Beyond the warm sunlit waters of tropical reefs, coral thrives in cool water—in the fjords of Norway, off the Aleutian Islands, in the Strait of Gibraltar, off the coasts of Florida and Ireland, in the sea near the Philippines. Cold-water coral lives as individual animals, small colonies, and giant banks that shelter thousands of other species. It grows on the continental shelf, in deep underwater canyons, and around seamounts. It grows in the dark and cold of the abyss, where cold water is saturated in calcium carbonate and where swiftly moving currents carry in nutrients and food. Anchored in the deep water, corals eat detritus from floating plants, zooplankton on their daily commute between depths and surface, and copepods wintering in the deep sea.

Deep-sea coral has endured for hundreds and, in some cases, thousands of years. Now both shallow and deepwater coral are under duress. The stresses are many.

Fishing a Reef

The deepwater coral, *Lophelia pertusa*, was once abundant in rich fishing grounds along the edge of Norway's continental shelf. Bottom trawlers crushed or buried between 30 and 50 percent of it. Banks of the white ivory tree coral, *Oculina varicosa*, live along the continental shelf off the Florida coast. Shrimp and fish trawling, scallop dredging, and depth charges released by the U.S. military searching for German submarines during World War II reduced much of this reef to rubble. Fisheries managers banned bottom trawling to protect the healthy thickets that remain, and they now require vessel tracking systems to reduce poaching in the protected *Oculina* reserve, whose coral is between 1,000 and 1,500 years old.

Fisheries managers also banned trawling in large areas of seafloor around the Aleutian Islands to protect deepwater coral gardens. Other countries are establishing reserves as well, but much of the sea's deepwater coral is still unmapped and unprotected. There is little time to lose. Trawling causes damage that will take hundreds, perhaps thousands, of years to repair.

Large fish disappeared from many warm, shallow water reefs long before scientists began keeping detailed records. A few remote, isolated, and relatively undisturbed reefs shine a light on that distant time, on a fullness that once was and still can be. In the Pacific's remote Line Islands, sharks make up 74 percent of the biomass on the uninhabited Kingman atoll but are virtually absent on heavily fished Kiritimati. On undisturbed reefs, marine food webs are inverted: the reefs are dominated not by tiny, colorful fish darting about the reef, but by large, cruising sharks. Historical accounts of Kiritimati describe sharks so numerous and aggressive they attacked the oars and rudders of boats. Sharks were abundant on Kiritimati as recently as 1997. In less than a decade of fishing, these sharks were gone.

Both deep- and shallow-water corals are vulnerable to destructive fishing practices. Left: Deepwater coral (Lophelia pertusa) *from the Trondheims Fjord, Norway. Right: Ornate butterflyfish* (Chaetodon ornatissimus) *on the heavily fished Kiritimati reef in the Pacific's Line Islands.*

But for these historical accounts and surveys of remote reefs, a marine food web top-heavy with sharks is one we would not know. Large sharks and jacks dominate reefs of the Northwestern Hawaiian Islands. They are bigger and more numerous than apex predators in the more urbanized and heavily fished reefs off the main Hawaiian Islands. Giant trevally jacks, Galápagos reef sharks, gray reef sharks, and other fishes at the top of the reef food web make up more than half the fish biomass on the remote reefs, but less than 3 percent on the main Hawaiian Islands reefs. It is a stunning difference, marking both the magnitude of loss resulting from intense exploitation and the possibility of renewal. In 2006 the Papahānaumokuākea Marine National Monument was established to further protect the roughly 140,000-square-mile (360,000-km²) area of the remote Northwestern Hawaiian Islands, one of the world's few remaining relatively pristine ecosystems.

Closing reefs to fishing, and prohibiting entry to fishing boats, can provide sanctuary for large fish and restore depleted populations. Little of the ocean—less than 1 percent—is so designated. Protecting between 20 and 30 percent of the sea's habitats in strictly enforced marine reserves is projected to cost between $5 billion and $19 billion per year, less than government subsidies to industrialized fishing fleets, whose capacity to remove fish is far greater than what the sea can sustain.

Gray reef sharks (Carcharhinus amblyrhynchus) *patrol Kingman Atoll in the Pacific's Line Islands. Sharks are the most abundant predators on this remote and undisturbed reef.*

The Waters Nearby

The Papahānaumokuākea Marine National Monument protects the remote Northwestern Hawaiian Islands, one of the world's few remaining relatively pristine ecosystems. Left: A coral complex (Montipora glabellata, Porites lobata, Pocillopora sp.) on Midway Atoll. Right: Red-footed boobies (Sula sula) in courtship on Green Island, Kure Atoll.

Coral reefs and humans share the edge of the sea with increasingly unhappy result. Years of deforestation, coastal building, and agriculture are releasing silt, nutrient-rich fertilizers, and wastewater into the sea, degrading the warm, clear water. Silt chokes coral. Nutrient-rich runoff fuels algal blooms, turning the water murky and blocking sunlight needed by corals' photosynthesizing tenants. New and deadly coral diseases are ravaging vast swaths of reef. The assault is killing tall, branching, majestic elkhorn and staghorn coral that sustained Caribbean reefs for thousands of years. Both elkhorn and staghorn coral are now on the U.S. endangered species list.

The combination of disease, pollution, and overfishing takes a toll as once vibrant reef communities fall apart. On Caribbean reefs, the disappearance of top predators and herbivorous fish, taken either for human consumption or for aquariums, left sea urchins to graze the reefs. Reduced redundancy within the ecosystem made reefs increasingly vulnerable. In 1983 long-spined, black Caribbean sea urchins (*Diadema antillarum*), important reef grazers, were suddenly killed by a fatal pathogen. With fewer herbivores to graze the reef, slow-growing coral, already weakened by disease and hurricanes, gave way to fast-growing seaweed.

Time passes, and threats facing coral reefs grow more complex. Tightly managed marine reserves can protect reefs from overfishing and destruction by ship anchors. Water degradation, whose source may be farther away, requires more complex and costly remediation: restoration of marshes, mangroves, and sea-grass beds that filter water and retain silt; sophisticated wastewater and stormwater treatment; less wasteful agricultural practices; and more restrictions on coastal development. These solutions, though more geographically widespread, are still local. They are being implemented in the Florida Keys National Marine Sanctuary to restore water quality over the reefs.

See Map 9, Marine Protected Areas, p. 269.

Corals feel the heat as earth's ocean warms. Pink coral (Acropora sp.) with bleached tips, Papua New Guinea. Right: Partially bleached polyps of daisy coral (Goniopora sp.) from Bunaken, Sulawesi, Indonesia.

Warming Sea

Coral reefs throughout the world, even those in pristine, remote areas, are now beset by a third host of problems whose sources and solutions are global. Large patches of ghostly white reef are appearing in tropical waters everywhere: along Australia's Great Barrier Reef, in the Caribbean, in the coral islands of the Seychelles. Photosynthetic algae living inside corals give them color and life. When the water is unusually warm for three or four weeks, corals expel their life-giving algae, and the reef bleaches. If water warms sufficiently, corals die. It doesn't take much, just 2 or 3 degrees Fahrenheit (1 or 2°C) above the seasonal high.

Only after millions of years did a tiny fraction of the earth's marine and terrestrial plants become vast oil reserves and coal basins. Today emissions from coal-burning power plants and automobiles quickly return that carbon to the atmosphere, warming earth's atmosphere and ocean to levels not seen in the last 650,000 years. Greenhouse gas emissions from human energy consumption rose by 70 percent between 1970 and 2004, and are continuing to rise: eleven of the last twelve years have been the warmest since the beginning of the Industrial Revolution. Corals feel the heat.

Unusually warm water in the Indian Ocean in 1998 triggered mass bleaching in the Seychelles. Seven years later, most of the reef had not recovered. The barrier reef off the coast of Belize bleached in the record warm waters of the Caribbean in 1998. Reefs in some areas lost their coral, a collapse unseen for at least three thousand years. For many Caribbean reefs, 2005 was worse. Australia's Great Barrier Reef has bleached eight times since 1979; the last bleaching affected more than half the reef.

Rising sea temperatures and repeated bleaching don't bode well for today's reefs. Neither do increases in carbon dioxide. Half our greenhouse gas emissions are lofted into the atmosphere, and a third are absorbed by the sea, rendering the water more acidic, less saturated in the carbonate corals need to build their skeletons. In increasingly acidic water, coral may cease to grow, eventually eroding and turning to rubble. The Intergovernmental Panel on Climate Change projects that by 2050, reef coral could become rare.

Safe Harbor

A reef is a crowded tapestry whose every nook and cranny teems with life. Thousands of schooling fish nibble coral and algae, dazzling the reef with color: neon blues, brilliant yellows, blood reds. It is a world both fiercely competitive, as corals vying for limited space turn deadly poisons upon each other, and cooperative, as nocturnal fish share a dwelling with those that feed by day, or as striped shrimpfish hang vertically, sheltered in the spines of sea urchins. A world without coral reefs would be a world sadly diminished.

Approximately 20 percent of the world's coral reefs have been lost in the last twenty to thirty years, and 60 percent may disappear in the next few decades. Controlling fossil fuel emissions, restoring water quality, and preventing overfishing can help protect coral reefs.

The loss of coral reefs is a loss of sustenance and livelihood, protection and defense against hurricanes and tsunamis, and unparalleled beauty and diversity of life. One species, *Homo sapiens*, may be on the edge of eliminating perhaps one-quarter of the sea's diversity. The fossil record, with its long perspective that carries over eons, points to the resiliency of reefs. They have survived, in one form or another, for three and a half billion years. During the Permian extinction, when earth's reefs had dissolved, a sea anemone without a shell somehow survived in the carbon dioxide–rich ocean, evolving into the coral we know today. Scientists drilling into rubble in deep water off the coast of Ireland found a mound of coral some two and a half million years old. It had lived for almost a million years, until advancing glaciers chilled its home. When the ice receded, 700,000 years later, it began growing again. This time, it grew for 500,000 years, until the current shifted, taking away food and nutrients.

During the Permian extinction, when earth's reefs had dissolved, a sea anemone without a shell somehow survived, evolving into the coral we know today. Left: Jewel anemones, Tasmania, Australia, closely related to coral, do not form a skeletal structure. Right: Hard coral (Mycedium sp.) showing individual polyps retracted, Palau, Micronesia.

Throughout earth's history, the sea has refilled with life. If a mass extinction of coral occurs on our watch, though, we may not be here for the recovery. Recovery from the Permian extinction took several million years, recovery from the Cretaceous extinction several hundred thousand years. For *Homo sapiens*, whose track record of endurance has yet to be proven, even a few hundred thousand years is a long time to wait.

A pink anemonefish (Amphiprion perideraion) finds refuge within the stinging tentacles of an anemone on a reef in Fiji.

9 Far Ends of Earth

Penguins live in one of the world's harshest environments.
Here, Adélie penguins (*Pygoscelis adeliae*) brave the freezing winds at Brown Bluff on the Antarctic Peninsula.

As spring comes to the Arctic and the winter sea ice retreats, hundreds of narwhals begin swimming north through Canada's Baffin Bay, following cracks and channels opening in the melting ice. They are heading for summer feeding grounds along the ice edge in the fjords and inlets of northwest Greenland and Baffin Bay. Their extraordinary tusks, up to 8 feet (2.4 m) long, were once believed to be the horns of unicorns. Unlike other mammal teeth, they contain millions of nerve endings close to the outer surface. The tusks, exquisitely sensitive to shifts in salinity and particles in the water, may help narwhals survive in ice-choked seas, guiding them toward polynyas, openings in the ice from which they breathe, and toward Arctic cod on which they feed.

As the sun rises higher in Arctic skies, the days grow shorter at the opposite end of the earth. Throughout the Antarctic summer, Adélie penguins bred and raised their young on the bleak, wind-swept coast of one of the world's harshest environments. As winter approaches and the sea freezes, penguins begin their annual migration onto the expanding pack ice. Walking or sliding on their bellies across the floes, then plunging into the water and shooting out onto the next floe, they follow the ice edge, where their food lies, as it tracks north. They will remain among the ice floes for the next nine months.

The Arctic is a frozen sea surrounded by land, and the Antarctic a frozen land surrounded by sea. Despite the dark, sunless winter, these waters support an abundance of life, adapted to bitter cold. The rhythms of life at the far ends of the earth follow the rise and fall of the sun, the advance and retreat of sea ice. In many ways, life at earth's poles is life on the edge.

Cold and Ice

Food webs at the poles are powered by a burst of life in the short summers, when the sun shines all day and night, and when the tiniest algae and the largest whale are buffered from the cold. Phytoplankton live in brine-filled interstices within the ice. They are grazed by copepods, amphipods, and euphausiids, whose slow metabolisms and rich stores of lipids carry them through dark, meager winters. In the nutrient-rich waters of Antarctic seas, these crustaceans and their predators (small fish) provide protein for larger seabirds, seals, large fish, and whales. Filtering diatoms from the water or scraping algae off ice, krill can grow to 2.4 inches (6 cm). Their numbers reach high into the millions. Come winter they sink into the depths and fast for months until the sun returns.

Larger animals have evolved a myriad of strategies to deal with cold. Arctic cod and Antarctic silverfish swim in icy 28.6 degrees Fahrenheit (-1.9°C) seawater. Antifreeze keeps their blood liquid. Emperor penguins nest on fast ice in the deep Antarctic winter, when temperatures plunge to -94 degrees Fahrenheit (-70°C). Huddling together in freezing winds and raging blizzards, they forgo territoriality, taking turns in the warm shelter of the huddle. Nursing baby seals drink milk that may be as much as 50 percent fat, enabling them to quickly grow an insulating layer of blubber.

During the summer, humpback whales feed in rich Arctic and Antarctic seas, gulping mouthfuls of krill-laden water or working collectively, blowing a curtain of bubbles that traps schooling fish. When the light fades, the temperature plummets, and food becomes scarce, the humpbacks leave, migrating long distances to their breeding grounds in warmer waters off the West Indies, Hawaii, Mexico, and Colombia. Arctic terns and mottled petrels make the longest known annual migrations of any bird, flying from one end of earth to the other and back again, following the sun, living in an endless summer.

Polar animals have evolved to deal with the bitter cold. Opposite above: Emperor penguins (Aptenodytes forsteri) *huddle on the ice of the Weddell Sea, Antarctica. Opposite below: Blubber helps insulate this Weddell seal* (Leptonychotes weddellii) *in Neko Harbour on the Antarctic Peninsula.*

The life cycles of many polar animals are tied to the annual advance and retreat of sea ice. In the Arctic, walruses ride the ice most of the year. Moving floes carry them over feeding grounds in the shallow waters of the continental shelf. Diving to the bottom, they stir it with their bristles, excavating thousands of clams each day. Ringed seals make their homes on the ice, clawing breathing holes as the ice grows thicker. They keep the holes open throughout the winter, building lairs in the snow above them. Ringed seals give birth in the lairs, where their nursing pups will be hidden from predators and protected from bitter winds until the sunlight returns and they've grown. Their adaptations to the ice have served them well: ringed seals, numbering in the millions, are the most abundant seals in the Arctic. Polar bears too depend on the ice, where they hunt ringed seals.

In the Antarctic, emperor penguins walk 60 miles (100 km) away from the open water of coastal polynyas to nest. Their long journeys back and forth across the ice to feed themselves and bring food to their chicks grow shorter as the ice melts in the spring, and open water advances toward the nesting colony. By the time the chicks are ready to take their first plunge into the sea, it is nearby.

Of Ice and Men in the Northwest Passage

For 400 years, European explorers have sought passage from the Atlantic to the Pacific across frozen Arctic seas. Most of their efforts failed miserably. In 1578 Martin Frobisher, though outfitted with fifteen ships, was distracted and bankrupted by a fruitless search for gold on Baffin Island. The more focused John Davis made three voyages to the Arctic, where he described abundant game and fish, and drew extensive maps. He was persuaded of the existence of the Northwest Passage but stymied by pack ice in Baffin Bay. Henry Hudson's trip ended in mutiny: Hudson died, and the starving mutineers lived on candle grease on their journey home. In 1615 William Baffin circumnavigated Baffin Bay, but ice blocked the way west, and he concluded that the passage did not exist.

Two centuries later, John Ross made the same circle around Baffin Bay. Ice turned him back as well. In 1845, provisioned with three years' worth of lemon juice, flour, beef, chocolate, and tobacco, and two ships, *Erebus* and *Terror*, whose double-planked hulls were designed especially for ice, John Franklin attempted to sail west from Baffin Bay through Barrow Strait into the Arctic. He and his entire crew perished. In 1906 Roald Amundsen, in a refitted herring boat, made it through. It was a circuitous journey around the islands of the Canadian Arctic, and a long one. Amundsen and his crew were locked in by ice for two winters before completing the voyage.

In September 2007, his transit would have been swift. Ice that had persisted throughout previous summers disappeared, opening the fabled shortcut between the Atlantic and the Pacific. Satellites passing over the Arctic captured images of snow-covered islands, clear water, melting blocks of older ice, and swirls of new ice forming as summer ended. It was a momentary glimpse into the Arctic summers' future, as greenhouse gases from fossil fuel emissions continue to warm the earth. Within a few weeks, the daylight faded, the channels iced over, and the Northwest Passage closed.

While ships generally require sturdy icebreaking equipment to cross the Northwest Passage, other travelers make the journey on their own. Diatoms from the Pacific Ocean, *Neodenticula seminae*, now bloom in the Atlantic. Scientists believe that, in the spring of 1998, they may have slipped through the Bering Strait into the Arctic, to ride a pulse of meltwater east during the summer. Then they may have drifted down through Baffin Bay into the Atlantic, taking up residence south of Greenland in the Labrador Sea. They have since expanded their range, blooming in the Irminger Sea off the coast of Iceland and in the Gulf of St. Lawrence. *Neodenticula seminae* is an important foundation of Pacific food webs. What role it will play in the Atlantic as Arctic ice continues to thin, further joining Atlantic and Pacific waters, remains to be seen.

In September 2007 the fabled Northwest Passage briefly opened. Opposite: Ice floes in the Gulf of Boothia, approaching Fury and Hecla Strait, off Baffin Island, Canada.

Below: Murre eggs collected by Inuit hunters on Bylot Island, Baffin Bay.

Plumbing the Depths

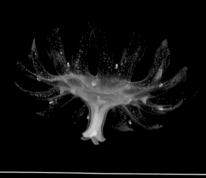

The poles haven't always been so harsh. Cores of sediment drilled from the Arctic, not far from the North Pole, reveal a balmy, freshwater past. Fifty-five million years ago, water temperatures in the Arctic were a warm 73 degrees Fahrenheit (23°C).

Fifty million years ago, the Arctic was a landlocked sea, filled with floating freshwater ferns. As time passed, drifting continents would change its nature, turning it from brackish estuary to salty ocean. As Greenland and North America drifted away from Europe, the Fram Strait separating them widened and deepened, and Atlantic seawater flowed into the Arctic.

It isn't easy taking stock of harsh waters. In the Arctic, three ships equipped with ice-breakers were required to allow scientists to drill successfully for cores amid the moving ice sheets. Exploration of the icy, stormy Antarctic is challenging as well, but as in the Arctic, each exploration yields new insight into a remote world. In the deep waters of the Weddell Sea, scientists recently found more than 700 new species of tiny animals: carnivorous sponges, crustaceans, mollusks, and swimming worms. This unexpected and rich diversity will help scientists understand how the advance and retreat of ice have altered climate and steered evolution in the deep sea.

Melting

In the Arctic, sea ice expands in the winter and shrinks in the summer. Today greenhouse gas emissions from humanity's increasing use of fossil fuels are warming the Arctic, accelerating the retreat of summer sea ice. At the end of the summer melt, Arctic sea ice is at its minimum. In recent decades, this minimum ice cover has been rapidly shrinking. Between the time when satellites first began measuring it in 1979 and when the Northwest Passage opened in the fall of 2007, it had shrunk by 40 percent.

Thick floes of multiyear ice are also dwindling. In the 1980s, 25 percent of the ice in the central Arctic was at least nine years old. That ice has disappeared. The loss is swift: between March 2005 and March 2007, the extent of year-round ice cover in the Arctic dropped by 23 percent. As Arctic ice becomes younger and thinner, less will survive the season of summer melting. The time is soon coming, perhaps in the next few decades, when much of the Arctic will be free of ice in the summer. The Arctic, icy and remote, a place apart, has long resisted the advance of humans. Now it is yielding.

Animals both large and small live in the fertile waters off Antarctica. Opposite top: From South Georgia, a Weddell seal (Leptonychotes weddellii) *resting on rocks. From the Weddell Sea, scyphomedusa larva (middle) and hydromedusa* (Calycopsis borchgevinki) *with extended tentacles (bottom).*

Below: Iceberg in Tracy Arm Fjord, Southeast Alaska. Thick glaciers are melting in the Arctic and Antarctica and throughout the world.

See Map 18, Arctic Sea Ice Concentration, p. 274.

Feeling the Heat

As the climate warms, some communities grow, and others fade. In the icy northern Bering Sea, the summer bloom of phytoplankton once supported a rich community of bottom-dwelling clams and mussels that fed gray whales, spectacled eider ducks, and walruses. As the ice cover thins and retreats, zooplankton grow fat on floating plants, and large numbers of pollock and pink salmon stream in to eat them. Bottom-dwelling animals, no longer supplied with carbon from plant communities living on the underside of the ice, decline.

Whale, duck, and walrus move north with the receding ice, and as the ice edge moves off the continental shelf into deeper water, walrus may suffer. Shallow water divers, they may become separated from the clam beds where they feed. One recent summer, when the water was unusually warm and the ice had retreated, scientists observed walrus pups in open water 9,800 feet (3,000 m) deep, separated from their mothers.

A walrus (Odobenus rosmarus) herd gathers on an ice floe. Walrus live much of the year on Arctic ice, from where they dive to the shallow bottom to feed on clams.

Polar bears (Ursus maritimus) on ice off northern Alaska. Sea ice is polar bears' lifeline. As the ice melts, their future is uncertain.

High in the northern sky, Ursa Major, the constellation of the Great Bear, rotates around Polaris, the North Star. Below the stars, on the sea ice, lives one of the Arctic's best known and most beloved animals, the polar bear. To grow, reproduce, and feed their young, polar bears, positioned at the top of Arctic food webs, depend on a diet of blubber-rich seals living on the ice. With their keen sense of smell, polar bears sniff out seals hidden in their snowy dens. Protected against the cold with a layer of blubber 4 inches (10 cm) thick and two coats of fur, they wait by openings in the ice for hours, or days if necessary, until a seal comes up to breathe.

As the Arctic warms, ice floes break up three weeks earlier in western Hudson Bay, forcing the bears ashore, where they fast until the ice returns. The bears are growing thin and hungry, and their numbers are dropping. Where stable sea ice recedes in the Beaufort Sea off Alaska, bears have left the ice, where their food is, to build their dens on land. Polar bears are strong swimmers, but in the Beaufort Sea, as the amount of open water increases, more bears drown. Their future is fraught with uncertainty: the United States Geological Survey projects that continued melting of sea ice would, by midcentury, reduce the number of earth's polar bears by two-thirds.

The West Antarctic Peninsula has warmed—2–3 degrees Fahrenheit (5–6°C) in the last fifty years, a faster rate than anywhere else on earth—and penguins are on the march. Sea ice on the peninsula, appearing later in the fall, melting away earlier in the spring, has, in the last twenty-six years, shrunk by 40 percent. As the ice disappears, chinstrap and gentoo penguins are extending their range from more northerly islands while Adélies are abandoning their homes, moving south toward the Antarctic continent. On Ross Island, Adélie and emperor penguins were forced on long detours to their nesting areas when an iceberg, the largest ever recorded, calved from the Ross Ice Shelf. Many penguins were trapped in the shifting ice. At the Beaufort Island emperor colony, sea ice disintegrated early, separating parents from their chicks before they had fledged. In both emperor colonies, populations plummeted.

As the West Antarctic Peninsula has warmed, chinstrap penguins (Pygoscelis antarctica), below, have extended their range from more northerly islands, while Adélies abandon their homes and move south.

A Culture at Risk

In the spring, narwhals (Monodon monoceros) follow the retreating ice to Arctic Bay off Admiralty Inlet, Baffin Island, to feed on cod.

Indigenous peoples have inhabited the Arctic for thousands of years, their sustenance and cultural, social, and spiritual identity intimately connected to its frozen land and icy ocean, to the hunting of seals, walruses, whales, and caribou, and to fishing and herding domesticated reindeer. As the sea ice retreats, knowledge accumulated by Arctic peoples over generations provides understanding and insight into the uncertainties created by the changing and unpredictable environment.

Inuvialuit (Inuit of the Canadian western Arctic) living in Sachs Harbour on Banks Island report fewer polar bears and skinnier seals as a consequence of earlier breakup and later freezing of sea ice. Thinner ice, less easy to cross safely and predictably, increases the hazards of traveling and hunting during the winter season. In Barrow, Alaska, early breakup exposes Iñupiat whalers to greater risks on shorefast ice during their spring whaling. Heavy storms triggered by the retreating ice and subsequent erosion are forcing residents of Shishmaref, Alaska, to evacuate their coastal village. Melting ice not only erodes land, but also puts new pressure on a way of life that has sustained the Inuit for centuries. They have successfully adapted to changing climate in the past. Now their future depends on adapting to what may be the greatest environmental changes they have ever experienced.

Beyond the Ice

Crabeater seals (Lobodon carcinophaga) *congregate at the edge of an iceberg in Pleaneau Bay, Antarctica. Despite their name, the diet of crabeater seals consists almost entirely of krill, which they sieve through specialized, multilobed teeth.*

Antarctica is a white, icy continent circled by a green, fertile sea. Millions of Weddell, crabeater, and leopard seals feed in its rich waters, along with blue, humpback, minke, fin, and killer whales. In many of Antarctica's waters, humans have made another mark, reconfiguring layers of food webs, removing animals large and small. We have taken more than one million whales from this distant sea; they have yet to return. We have taken millions of tons of fish—shallow-water rock cod and then, as shallow-water fisheries collapsed, deeper-dwelling, long-lived toothfish (also known as Chilean sea bass). And now we are considering removing additional amounts of krill, the staff of life for many Antarctic marine animals. Watery, diaphanous salps have replaced krill in some Antarctic waters. As a source of protein, they are no substitute for meaty krill.

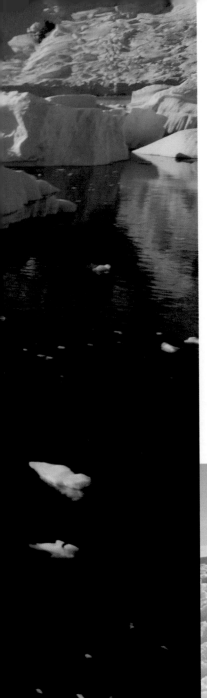

Iron sometimes fertilizes the sea. When the wind blows iron-rich dust into a sea rich in other nutrients, phytoplankton bloom, drawing carbon dioxide from the atmosphere. Today entrepreneurs are preparing to seed the sea with iron, hoping for a similar result that will mitigate the warming effects of fossil fuel emissions. There may be consequences, both near and distant, anticipated and unanticipated. Copepods and krill may graze the meadows, recycling carbon, exhaling carbon dioxide as they breathe. Decomposing plants may deprive deep waters and their residents of oxygen, while releasing other greenhouse gases such as nitrous oxide and methane. By triggering blooms of selected species, iron fertilization restructures the base of marine food webs; the changes may reverberate throughout the entire web.

The waters of the sea are connected; coercing the sea to bloom in one part of the ocean may leave it barren in another. Farmers fertilizing fields in the Mississippi River watershed fertilize the Gulf of Mexico, setting off phytoplankton blooms that each summer lead to the creation of a dead zone the size of Massachusetts. The Intergovernmental Panel on Climate Change has found that fertilization of the ocean is "largely speculative and unproven, and with the risk of unknown side-effects." Oceanic food webs are finely balanced; we are far from understanding the possible reverberations of dumping iron into the sea.

Orcas (Orcinus orca) *surface in McMurdo Sound, Antarctica. Killer whales such as these live throughout the world ocean at the top of marine food webs.*

Reverberations

Cartographers are redrawing maps of coastal Greenland. As a land buried in ice begins to thaw, what were once thought to be peninsulas jutting out into the sea are now revealed as islands. Thick glaciers are melting, both in Greenland and Antarctica. Torrents of running water plunge into crevasses within the ice, lubricating the glaciers, accelerating their slide to the sea. In the last five to ten years, as the Jakobshavn Isbrae and Kangerdlugssuaq glaciers make their way to the sea, their speed has doubled. Where glaciers meet the sea, they are nibbled by warming water, and big chunks break off.

As the fossil fuel energy that powered contemporary civilization warms the far reaches of earth, the poles are beginning to lose their glaciers, and white lands are turning green. Earth's continents are joined by flowing water. As we warm earth's polar realm, water from these melting glaciers will touch our shores. It has in the past. By 2100, and perhaps earlier, Greenland will be as warm as it was 130,000 years ago. Back then, portions of Greenland and Antarctica melted. The sea rose 13 to 20 feet (4 to 6 m), and it rose quickly, perhaps more than 3.3 feet (1 m) in a hundred years.

Tabular iceberg in Antarctica's Bransfield Strait. Icebergs as large as the state of Connecticut have broken off Antarctica and floated into the sea.

Today the sea is already rising more quickly than we had anticipated, and our calculations have not included rapid disintegration of Greenland and Antarctica's ice sheets. Unless we substantially and immediately reduce greenhouse gas emissions, a flood we thought would come a thousand years from now may come much sooner, within the lifetimes of our children and grandchildren. Innocently, then willfully nudging the atmosphere, we may create an earth where snow and ice will cease to have palpable meaning. The far ends of earth may lose their identities as remote and icy wildernesses where ice crystals hang in the air, where the snowy silence stretches forever, and where narwhals follow the moving ice edge. If polar glaciers turn to water, sea level across the earth will dramatically rise, and changes in a realm remote and distant will reverberate nearby.

Where Rivers Meet the Sea

In estuaries, rivers return to the sea. Here, a creek winds its way to the ocean near Townsville, Queensland, Australia.

A small woodland creek opens onto a salty mudflat flooded by the tide. A river rises high in the mountains, travels across a continent, then descends onto a coastal plain, broadening, slowing, and turning salty as it meets an incoming tide. In estuaries, rivers return to the sea. Carved by rivers or glaciers and flooded by a rising sea, or fashioned as currents build a sandbar at the mouth of a river, they are inhabited by plants and animals adapted to the stress of salinities that change with the tides. Their sheltered waters serve as nurseries for young fish. Great flocks of migrating birds pause on their journey to forage in the rich mud. Liminal places belonging to both continent and ocean, estuaries are among the most beautiful and fragile places on earth, and the most touched by people, some of whose largest cities— New York, Shanghai, Buenos Aires, Kolkata (formerly Calcutta)— lie within them.

Newcomers

Chincoteague Island, Virginia. Chincoteague and Chesapeake Bay were once covered by a shallow sea reaching inland almost as far as Washington, D.C.

In the long reach of geologic time, today's estuaries are recent arrivals. Between 20 and 10 million years ago, the area that would become one of America's largest estuaries, Chesapeake Bay, was covered by a shallow sea reaching inland as far as Washington, D.C. Tidal marshes, sandy beaches, and bald cypress swamps bordered this sea. Porpoises and dolphins, sharks and rays, sea cows and sea turtles swam in its waters. There were seabirds: puffins, fulmars, and gannets. Whales came to calve, and earth's largest predatory shark, *Carcharodon megalodon*, passed through. Its serrated teeth were up to 7 inches (18 cm) long, and its body extended at least 49 feet (15 m). This giant grew up to three times the length of today's great white sharks.

Eventually the water drained away. Eighteen thousand years ago, when the Northern Hemisphere was covered with ice, sea level dropped 330 feet (100 m), and rivers, among them the Susquehanna, cut their way across Maryland to the edge of the continental shelf. When the ice receded, the sea rose again, drowning the Susquehanna valley and creating Chesapeake Bay. The water is still rising.

Estuaries come in many shapes and sizes. During the ice ages, moving glaciers cut deep U-shaped valleys in the mountains of Norway and New Zealand. When the ice melted, between fifteen and ten thousand years ago, the sea rushed in, creating deep fjords—Geiranger Fjord in Norway, Milford and Doubtful sounds in New Zealand. These are estuaries, as are the warmer waters in Pamlico Sound, sheltered behind miles of North Carolina's barrier beaches, and the cold waters at the mouth of Washington State's Columbia River, an estuary so wide that Lewis and Clark and the Corps of Discovery, upon reaching a point some 20 miles (32 km) inland, mistakenly thought they had reached the sea. Over centuries and millennia, estuaries emerge and disappear as rivers change course, glaciers advance and recede, and sea level rises and falls. While they are here, they fill the mutable place between land and sea where boundaries are soft and blurred, and edges move with the tides.

Dawn breaks over Milford Sound and Mitre Peak, New Zealand. Advancing glaciers cut deep valleys that would become fjords when the ice retreated and the sea rushed in.

See Map 7, Salt Marshes, p. 268.

Rimmed by Salt Marsh

Rapidly changing temperature, salinity, and oxygen levels challenge those that would live in an estuary. Not many have adapted to its inconstant nature, but those that have, thrive. Cordgrass, *Spartina*, filling salt marshes within the highest reaches of the tide, excretes excess salt through special glands. The succulent sea pickle, *Salicornia*, holds water in fleshy stems. Oysters snap their shells closed when salinity drops, then reopen them when the tide returns. Their strategy has worked well. Chesapeake Bay, whose Native American name means "Great Shellfish Bay," was once crowded with oysters piled so high they posed a navigational hazard, and so numerous they could filter as much water as the bay holds—19 trillion gallons (72 trillion l)—in one week.

Horseshoe and blue crabs also thrived in the bay's changing water conditions. During the spring, when sun, moon, and earth are aligned, and the tide is at its monthly high, horseshoe crabs come into the bay to spawn, laying their eggs in the beach at the edge of the marsh. Two weeks later, when sun, moon, and earth are again aligned, and high tides are again at their highest, the eggs hatch, and the flooded larvae are carried out to sea. Blue crabs swim throughout the entire bay. Millions of female blue crabs migrate south every autumn to give birth in salty waters near the sea. The young then journey back up into the bay's increasingly fresh water, seeking shelter from predators in underwater grasses.

Baby sharks spend their youth in the calm, sheltered waters of Bulls Bay, in South Carolina's Cape Romain National Wildlife Refuge. In spring, blacktip, sandbar, dusky, hammerhead, and spinner sharks give birth in the warm, shallow water. After giving birth, the mothers leave, but the young stay in the estuary, where they are protected from predation and nourished by the water's abundant menhaden and shrimp.

Saltmarsh cordgrass (Spartina alterniflora) in Willapa Bay, Washington. Saltmarsh cordgrass is well adapted to the inconstant waters of estuaries. Right: A blue crab (Callinectes sapidus) snaps its claws in defense against a northern seahorse (Hippocampus erectus) off Kiptopeke in Cape Charles, Virginia.

See Map 8, Mangroves, p. 269.

Rimmed by Mangrove

Closer to the tropics, estuaries are fringed not with salt marsh, but with thick forests of mangrove. Excreting excess salt through their leaves, they too are well adapted to an environment washed by the tide. Stiltlike roots prop up red mangroves in soft, water-logged mud. Exposed to the air, the roots can breathe. The mud surrounding a black mangrove may be spiked by hundreds of pneumatophores, pencil-like breathing tubes sent up into the air from its roots. Adaptations of mangrove roots and leaves secure their lives in airless, salty mud. Adaptations of seeds secure their future. Seeds of red mangroves germinate right on the branches, where they grow heavy roots. Ballasted by their roots, seedlings float upright when they drop into the water. They ride the tides, brush against soft mud, and anchor. As they grow, the tangled, impenetrable roots of these pioneers trap sediment and extend the reach of dry land. When Marco Polo visited Palembang, Sumatra, in 1292, it was a thriving port city. Today it lies 31 miles (50 km) inland, on land built by mangroves advancing into the sea.

Mangroves are home to both terrestrial and marine life. Storks, egrets, flamingos, toucans, and spoonbills roost in the trees, while oysters, snails, and crabs nestle in the nooks and crannies of mangrove roots. Mudskippers are at home in both worlds, swimming when the tide is high, breathing through their gills. When the water ebbs, they breathe air, climbing onto mangrove roots or skipping through the mud on their pectoral fins. The trees and their inhabitants live together, providing for each other. Crabs find food and shelter amid the tree roots and, burrowing in the mud, aerate them.

Coral reef and mangrove forest nurture each other as well. Reefs break waves, providing calm water for mangroves. Mangroves filter nutrients and silt from rivers flowing into the sea, producing clear water needed by coral. Animals rely on both. Juvenile shrimp, mullet, and spiny lobster live in mangrove nurseries, then move out onto the reefs as adults. The numbers of snapper, bluestriped grunt, and rainbow parrotfish on coral reefs surge when mangrove nurseries are nearby.

Mangroves are well adapted to an environment washed by the tide. Here, a red mangrove in Tunicate Cove, Belize. Right: A saltwater crocodile amid mangrove roots in the Philippines.

Chesapeake and Beyond

Little by little, we have altered the rhythms of life in estuaries, pivotal places where land meets sea and where the world of water meets the needs and desires of humans. Gray whales, extirpated long ago, once swam along America's eastern seaboard, perhaps breeding in Chesapeake Bay. A partial skull found near the mouth of the bay signals their former presence. Thomas Hariot, representing Sir Walter Raleigh, visited Pamlico Sound in 1585 and found enormous fish: 2-foot (.6-m) -long herring, 5-foot (1.5-m) striped bass, and 10-foot (3-m) sturgeon. John White's drawings from that expedition show hammerhead sharks beneath Native American canoes. Fish were so plentiful that Native Americans stood in the shallows, spearing them with sharp poles. This fullness is now diminished.

Many are the strands in Chesapeake Bay food webs. Marsh and forest along the water's edge trap sediment and absorb nitrogen flowing into the bay. Oysters filter water, maintaining its clarity, helping sunlight reach meadows of underwater eelgrass. Eelgrass shelters juvenile blue crabs, breathes oxygen into the water, traps sediment, and removes nitrogen. All yield to human design. The community is stressed, and slowly the bay is dying. Oysters that gave Chesapeake Bay its name have long been taken. So few remain, they now take many months to filter a volume of water equal to that in the bay. The Chesapeake's famous blue crab, whose numbers have also plummeted, is not reproducing well; its eelgrass nursery has shrunk to a fraction of its former size. Strict regulation of the striped bass fishery has restored their numbers, but their health is now compromised by polluted water.

Animals frequent the edge between land and sea. Below left: Sanderlings (Calidris alba) in Chesapeake Bay. Right: A sika deer (Cervus nippon) in a marsh in Chincoteague, Virginia.

Nitrogen and phosphorous entering the Chesapeake—from lawn and farm fertilizers, from animal farms and sewage, and from automobile and power plant emissions—overwhelm the bay's ability to renew itself. These excess nutrients fuel blooms of algae that cloud the water and, as they decompose, remove its oxygen, thinning eelgrass

meadows and killing fish and oyster larvae. During the summer, when the bay suffocates, thousands of menhaden, perch, largemouth bass, and blue crabs die. Sturgeon, once prolific in the bay, now would have difficulty reproducing and maturing in the degraded water.

Losses like those in the Chesapeake and the Gulf of Mexico dead zone occur in estuaries across the world. Harmful blooms of algae also choke marine life at the mouth of the Chang Jiang (Yangtze) and in the Baltic, Adriatic, and North seas, fouling the air, rendering fish inedible, harming sea life. Reproduction is dramatically impaired in both male and female Atlantic croakers growing up in the oxygen-poor waters of Florida's Pensacola Bay. Humans have depleted coastal estuaries and seas of 90 percent of the large animals that once dwelled there, removing whale and sea otter, auk and egret, tuna and shark, salmon and sturgeon. We have removed 65 percent of the wetlands and undersea grasses. Now many of earth's estuaries have difficulty breathing.

Left: A brown bear (Ursus arctos) *runs across the sand in Katmai National Park, Alaska. Right: A brown pelican* (Pelecanus occidentalis) *feeds its chick on Shank's Island, Virginia.*

Wherever we live, we live in watersheds that drain to the sea. Forty percent of the human population lives near the edge of the ocean (within 60 miles or 100 km), and the rest lives near creeks or rivers that ultimately flow into the ocean. The watershed of the Mississippi River encompasses 40 percent of the lower forty-eight U.S. states. Farmers tending fields in Kansas and Nebraska also tend to the lives of shrimp in the Gulf of Mexico.

More than two-thirds of assessed estuaries in the United States show signs of eutrophication. This degradation is not inevitable. Reduction of fertilizers flowing into the Black Sea is restoring water quality in one of the world's largest dead zones. In the Chesapeake Bay and in estuaries throughout the world, controlling growth, upgrading sewage treatment plants, replenishing wetlands, underwater grasses, and oyster beds, reducing agricultural runoff, and restoring riparian woodlands together offer hope for repairing these fragile worlds.

Domestication of Estuaries and Coastal Waters

The number of invasive species in coastal waters is increasing. Above left: Lionfish (Pterois volitans), originally from the Indian Ocean and the Pacific, are increasing along the U.S. eastern seaboard. Above right: Caulerpa taxifolia now carpets the Mediterranean, crowding out native grasses and animals.

Humans shape earth's estuaries, by accident and by design. Some 110 million years ago, earth's continents were joined into one land surrounded by one sea. The continents will not join again for another 200 or 250 million years, but commerce and trade, originating in cities that grew where rivers meet the sea, already join estuaries separated by wide ocean basins. Merchant ships brought green crabs to America from Europe, and Viking ships took American clams to the Baltic Sea. Pacific shore crabs now wind their way through Atlantic coastal waters, decimating clam flats, and veined rapa whelks from the Sea of Japan spread through Chesapeake Bay, eating oysters and clams.

They could not swim these great distances from their original homes. In all likelihood, their larvae were stowaways in ballast water, perhaps in ships coming through the Panama Canal. Thousands of ships, discharging millions of gallons of ballast, mix waters from opposite sides of the ocean. Once, seeds carried on the wings of birds or lofted by winds, and larvae borne by wave and current, were the primary means by which plants and animals expanded their ranges. Now the passage of humans holds the greater influence.

Invasive species are on the move. In the Mediterranean, bright green seaweed, *Caulerpa taxifolia*, in all likelihood flushed from an aquarium, now carpets the seafloor, crowding out many native underwater grasses, invertebrates, and fish. Large red lionfish, *Pterois volitans*, with venomous spines, live in the Pacific and Indian oceans. Brought to the United States by the aquarium trade and then accidentally released into the sea, they have spread from Florida all the way to North Carolina, where, lacking predators, their increasing numbers are changing the face of reef communities.

Humans have always looked to the sea for sustenance, and as our needs expand, so do our expectations that the ocean will provide. We seek to capture energy from tides, waves, and icy methane hydrates buried on the continental shelf and to dispose of carbon dioxide emissions on or in the seabed. In estuaries, where cities cluster, we seek from the sea its most voluminous resource: water. Twenty percent of the world's population suffers from a scarcity of freshwater; 130 countries use some form of desalination. Water-starved countries across the world—Saudi Arabia and other Persian Gulf nations, North Africa, Australia, China, and the western United States—turn saltwater into fresh, then discharge the leftover brine, now twice as salty, back into the sea. We have little knowledge of how the delicate balance in estuaries and coastal waters and the life they hold will be altered by this infusion.

Across the earth, in the cold fjords of Norway and the mangrove forests of the tropics, estuaries are making room for salmon pens, oyster rafts, and shrimp farms. Salmon and shrimp farms, producing nitrogen waste to rival that in untreated human sewage, degrade the water. The deforestation rate of earth's mangroves in the last twenty years exceeds that of tropical rainforests. Thick, tangled mangroves dissipate the violence of storms and floods, and their quiet, protected waters provide nurseries for young wild fish. These benefits are lost when mangrove forests are cut for firewood and converted to cattle pastures and shrimp ponds. After five or ten years, if salt, waste, and chemicals accumulate in aquaculture ponds, they may be abandoned. Oyster and mussel cultivation is less deleterious to water quality. Farmed mollusks both contribute to the supply of seafood and clean the water, filtering it as they feed.

Below: Caribbean mangrove forest. Mangroves provide nurseries for young fish and filter nutrients and silt from rivers, producing the clear water essential to coral reefs.

Rising Water

Rivers wash mountains into the sea. The Ganges, Brahmaputra, and Meghna rivers carry soil and silt down from the Himalaya into the Bay of Bengal, building the world's largest river delta, where the largest remaining tract of mangroves grows. The rivers divide and divide again, and still they are as much as a mile (1.6 km) wide, braiding around low islands and mud banks. Half the Sundarbans, as this area is known, are under water. Land merges into sea, and sea into land, each day with the tides. Islands appear and disappear as monsoon floods ebb and flow. Dolphins swim in tidal rivers, and wild deer, boar, and macaques live in the forest, along with our only predator. Endangered Bengal tigers live in the Sundarbans, where at times they prey upon fishermen, woodcutters, and gatherers of wild honey. Protecting tigers helped protect mangroves, but now the water is rising.

Some 2,500 years ago, the ancient city of Herakleion flourished in the Nile Delta. A legendary and wealthy port at the mouth of the river, it rose to prominence and then vanished. In 1999 archaeologists found the lost city. Its walls and streets, and remnants of temples, columns, and statues, lie in the Mediterranean, in Abu Qir Bay, submerged under 16 to 23 feet (5 to 7 m) of water. Herakleion, built on the soft muds of a river delta, sank as the underlying and unconsolidated sediment collapsed and the sea level rose.

See Map 17, Sea Level Rise, p. 274.

The rising sea threatens many of the world's largest cities and its smallest island nations with flooding. Below: Dhaka. Bottom: Baa Atoll in the Maldives.

Safe harbor, navigable river inland, fertile delta, strategic location for commerce or territorial control: these are some of the many reasons cities grow at the mouths of rivers. They are increasingly unstable places. Jean-Baptiste Le Moyne, Sieur de Bienville, established the settlement of New Orleans on a patch of soggy Mississippi River delta in 1718. In 1719 it flooded. As the marshes were drained and the city grew, New Orleans began sinking under its own weight. It lies below sea level, virtually surrounded by water, in an area historically battered by storms. In 2005 Hurricane Katrina's 16-foot (5-m) storm surge flooded three-quarters of New Orleans, destroying entire neighborhoods and forcing the evacuation of over a million people.

Many of earth's river deltas are sinking, as thirsty cities pump groundwater and divert and dam rivers. More than 45,000 reservoirs and their associated dams trap river sediment upstream, starving deltas. Greenhouse gas emissions warm and expand the ocean, melting glaciers add more water, and sea level rises. As it does, millions of people living in the Nile, Chang Jiang (Yangtze), Mekong, Mississippi, and Ganges/Brahmaputra/Meghna deltas may lose their homes to the sea, and millions more will be flooded in storm surges spawned by hurricanes and typhoons. Bangladesh lies in the Ganges delta; by 2050, storm surges will touch almost 50 percent of the country. When the sea rises 18 inches (45 cm), three-quarters of the Sundarbans' mangrove forests will disappear. Sea-level rise threatens some of the world's largest coastal cities—Mumbai, Dhaka, and Shanghai—as well as the sustainability and sovereignty of small, low-lying island nations such as Tokelau, the Maldives, and Tuvalu.

A great blue heron (Ardea heridias) stands in the salt marsh of the Blackwater National Wildlife Refuge, Maryland.
If the land here continues to subside, and the sea continues to rise, most of the marsh may become open water by midcentury.

In Chesapeake Bay, the water rises. Islands once inhabited by settlers in the bay have disappeared, their farms and fields flooded. Bay steamers once called at Sharps Island, at the mouth of the Choptank River. Today only the lighthouse, standing askew in 10 to 13 feet (3 to 4 m) of water, is left. Tangier Island, in the middle of the bay, less than 6.5 feet (2 m) above the sea, is central to the history of bay watermen and the blue crab fishery. Now, at high tides, water flows into the island's low-lying marshes, prelude to a more lasting inundation.

Over millions of years, the birth of mountainous ridges beneath the sea has opened oceans, spilling water onto the continents. More recently, over thousands of years, sea level has ebbed and flowed with the advance and retreat of glaciers. Estuaries and river deltas have been inundated by the ocean, drained by advancing ice, and then flooded as the ice retreated. Now, in our time, we too are calling the rising water.

The sea is our lifeline, and we are its partner.

Our Water, Our World

Ancient seafloor built into continents holds a record of earth's history. So too does the ocean now record our own human imprint on the planet. Our mark is being recorded on the floor of the Mediterranean, where Herakleion and the lost cites of Egypt now lie submerged; on the floors of the Chesapeake Bay, the Baltic, and the Gulf of Mexico, whose sediments describe waters that each summer can no longer breathe; and along the world's continental shelves, where fossils will tell of the disappearance of the sea's large fish, whales, and coral reefs.

Is this the legacy that we, the last human species, will leave of our time on earth? Will the seabed record a time when the sea suffocated from our excesses? When our claim on earth's dwelling spaces forced so many of earth's other inhabitants to lose their homes? When the desires of so few were met at such great cost by so many?

We follow a path taken before. Millions of years of earth's history recorded on the seabed suggest that doublings of atmospheric carbon dioxide produce an average 5-degree Fahrenheit (2.8°C) increase in temperature, that life's diversity can be impoverished in a greenhouse world, and that mass extinction is often associated with rises in atmospheric carbon dioxide and acidic seas. Four out of five of earth's major mass extinctions were associated with rises in atmospheric carbon dioxide.

Our fossil fuel emissions, habitat destruction, overfishing, introduction of invasive species, and pollution may be triggering earth's largest extinction since the death of the dinosaurs. Mass extinctions diminish the diversity of life and alter the course of evolution. We are creating a world where dazzling diversity may be reduced to bland homogeneity; where fast-growing, opportunistic species may replace longer-lived, slower-growing large organisms; where large animals may no longer evolve; and where "hot spots" of evolution—tropical rainforests, coral reefs, mangroves—may disappear. In a mass extinction, many species die, and fewer new ones evolve. The aftermath can last for millions of years.

In the long reach of time, the sea is resilient and will endure. Life in the ocean has persisted through extreme conditions: scalding vent waters, icy freezes of advancing glaciers, corrosive waters of a greenhouse earth, and the violent, devastating upheaval of colliding continents. Its tenacity has been bolstered by one highly successful partnership after another. Early on, a protective membrane enclosed genetic material to build a living cell. Later, individual, independent cells joined into multicellular powerhouses to spawn the sea's first plants and animals. These innovative unions began at a cellular level, but over time they expanded into partnerships between organisms—symbiotic bacteria living in giant tubeworms at deep-sea hot springs, photosynthetic algae living in coral, bioluminescence-producing bacteria living in deep sea animals—and partnerships among communities.

The overarching partnership—enduring since the dawn of life and that works, seen and unseen, through every food web—is between life and the sea. Evolution depends on that partnership.

The genetic blueprint for a large animal existed long before the organism was built, and the capacity for fish to walk existed long before it was expressed. Only when the sea filled with air enough to breathe did animal life burst forth. Only when eroding mountains created floodplains could barren land become green and fish come ashore. Only after a sea change did *Australopithecus* begin the transformation to *Homo*.

Opening and closing ocean basins rebuild continents, and the remains of those that dwell in the ocean are recycled into seafloor, raised into mountains, and then washed back into the sea to nourish another generation of plants and animals. Flowing currents send rain to dry land and bring nutrients from the depths to build vast food webs and fertile fisheries. We are far removed from the invisible plants that constitute the sea's primary producers, but we cannot live without them. The sea is our lifeline, and we are its partner.

Our chapter in earth's history has yet to close. The last pages are not yet written. Now no part of earth's vast ocean is untouched by humans; on 41 percent of the sea, our mark is heavy.

Unlike the animals that came before us, we can choose the legacy we leave. Our choices will be felt for generations to come, and by all who dwell on earth.

We can choose a legacy of stewardship, choose to live by an ocean ethic, whose principles, derived from the nature of the sea itself, are these:

— that the sea is the source and sustainer of all life, including ours;

— that we are but one species among many;

— that earth history recorded in ancient seafloor illuminates our present and intimates a future;

— that earth's ocean waters are joined in a single, flowing sea;

— that wherever we live, however we live, we touch the sea and therefore share responsibility for its health and well-being.

We are a species blessed with language and awareness, knowledge and ingenuity, and an infinite capacity for imagination. May ours be a time remembered when we reconceived our partnership with the sea and restored its life-giving waters to health and wholeness. May ours be a time remembered when the sea's habitats and inhabitants, wherever they may be, found true sanctuary—in salt marshes and seamounts, coral reefs and continental shelves, shimmering surface and deep abyss. May ours be a time remembered when we embraced the sea as it embraces us.

Common brittle star (Ophiothrix fragilis) *from the North Sea.*

Philippine shrimp (Vir philippinensis) live and feed on bubble coral (Plerogyra sinuosa), Papua New Guinea.

Atlas

Note: Names printed in red on the maps are discussed in the text.

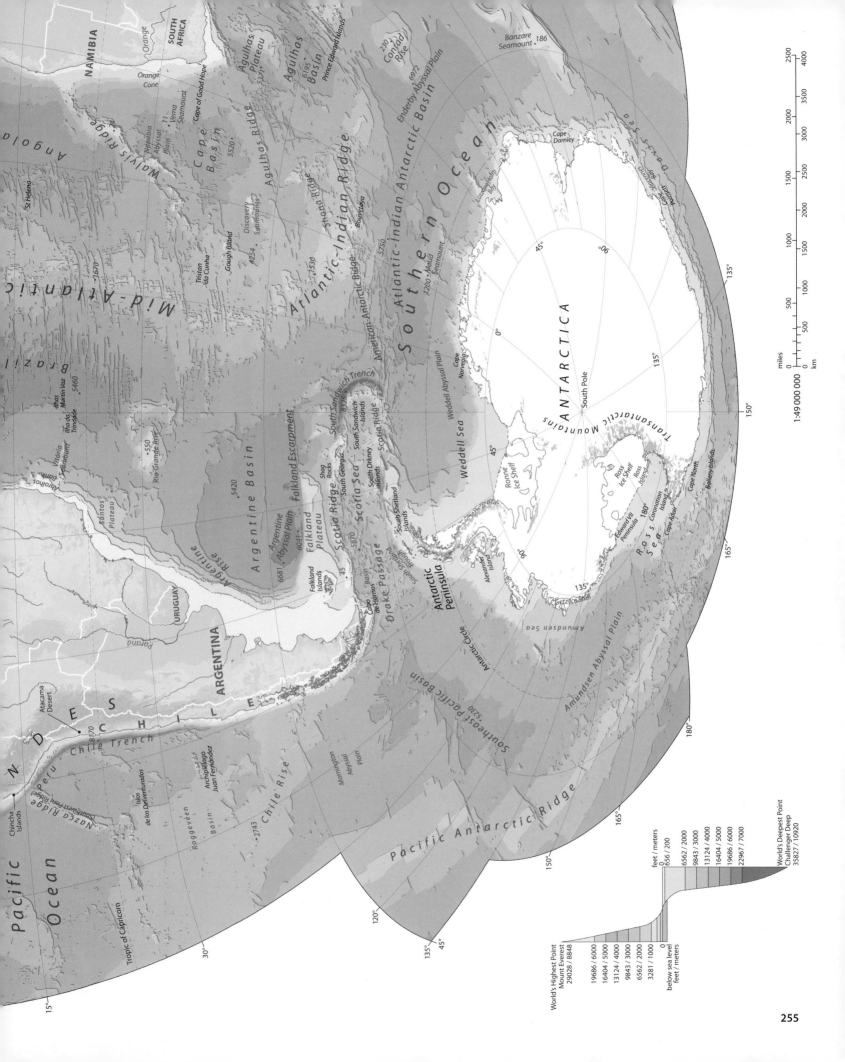

INDIAN OCEAN

Great Barrier Reef

New Guinea

Pacific Ocean

RUSSIAN FEDERATION

Lena

Ob'

Lake Baikal

Hokkaido

Honshu

Vityaz Depth 10,542

8412

South Honshu Ridge

Shikoku

Kyushu

Kyushu–Palau Ridge

Nansei-shoto

1404

8181

Japan Basin

3510

Sea of Japan (East Sea)

Korea Bay

Bo Hai

Yellow Sea

East China Sea

Taiwan

Taiwan Strait

Luzon Strait

Ryukyu

Batan Islands

Cape Engano

Luzon

Mount Pinatubo

Philippine Islands

Philippine Basin

6745

Philippine Trench 10,057

Mindanao

Palau 8054

AUSTRALIA

Halmahera

Seram

Laut Seram

Laut Banda

Celebes Sea 5484

Sulu Sea

Sulawesi

Laut Jawa

Laut Flores

Sumba

Java Ridge

Tambora

Timor

EAST TIMOR

Melville Island

Cape Leveque

Arafura Sea

Arafura Shelf

Torres Strait

66

Cape Arnhem

Gulf of Carpentaria

Timor Sea

I N D O N E S I A

Borneo

Palawan

Palawan Trough

Hainan

Gulf of Tongking

South China Sea

5560

SINGAPORE

Sunda Shelf

Bangka

Mui Ca Mau

Yellow

Yangtze

C H I N A

Mekong

Gulf of Thailand

THAILAND

22°

MALAYSIA

Sumatra

Strait of Malacca

Kepulauan Mentawai

Java

Java Trench (Sunda Trench) North

7145

Christmas Island

6360

Cocos Islands

Exmouth Plateau

Australian Basin

North West Cape

Shark Bay

1924

Shark Bay

Perth Basin

5746

Naturaliste

West Australian Basin

Investigator Ridge

East Indiaman Ridge

Broken Plateau

2067

7349

MYANMAR

Irrawaddy

Brahmaputra

Meghna River

BANGLADESH

Ganges

Ganges Cone

.3954

Bay of Bengal

Andaman Islands

Nicobar Islands

Andaman Basin 4267

2302

Cocos Basin

Ninetyeast Ridge

3745

HIMALAYA

Ganges

I N D I A

PAKISTAN

Indus

Indus Fan

Gulf of Khambhat

Mumbai

Arabian Basin

3694

Al Masirah

Arabian Sea

Laccadive Islands

Cape Comorin

Gulf of Mannar

Sri Lanka

SRI LANKA

4735

MALDIVES

Maldives

Chagos Trench

Chagos–Laccadive Ridge

5406

Diego Garcia

Chagos Archipelago

British Indian Ocean Territory

Mid-Indian Basin

5421

Mid-Indian Ridge

Carlsberg Ridge

1682

5803

1481

Somali Basin

5060

Socotra

Seychelles

Amirante Islands

Amirante Trench

5273

SEYCHELLES

Farquhar Islands

Aldabra Islands

Saya de Malha Bank

Mascarene Plateau

Mascarene Ridge

Cargados Carajos Islands

Agalega Islands

Ile Tromelin

Rodrigues Island

MAURITIUS

Mauritius

Reunion

Mascarene Basin

Mascarene Plain

Madagascar Basin

5408

Abra

2067

Aral Sea

Caspian Sea

1025

IRAN

OMAN

YEMEN

Strait of Hormuz

Gulf of Oman

The Gulf

Volga

Black Sea

2210

Tigris

Tell Leilan

Euphrates

Fertile Crescent

Mediterranean Sea

Dead Sea

3039

Red Sea

Gulf of Aden

SOMALIA

SUDAN

Tropic of Cancer

30°

15°

Awash River

Afar Depression

ETHIOPIA

Lake Turkana

KENYA

Olduvai Gorge

TANZANIA

Equator

Pemba Island

Zanzibar Island

Mafia Island

COMOROS

Njazidja

Nzwani

Comoro Islands

Moroni

Mayotte

Mozambique Channel

Ile Europa

Bassas da India

Juan de Nova

MADAGASCAR

Madagascar

Komati River

Tropic of Capricorn

MOZAMBIQUE

Mozam

60°

45°

30°

15°

0°

15°

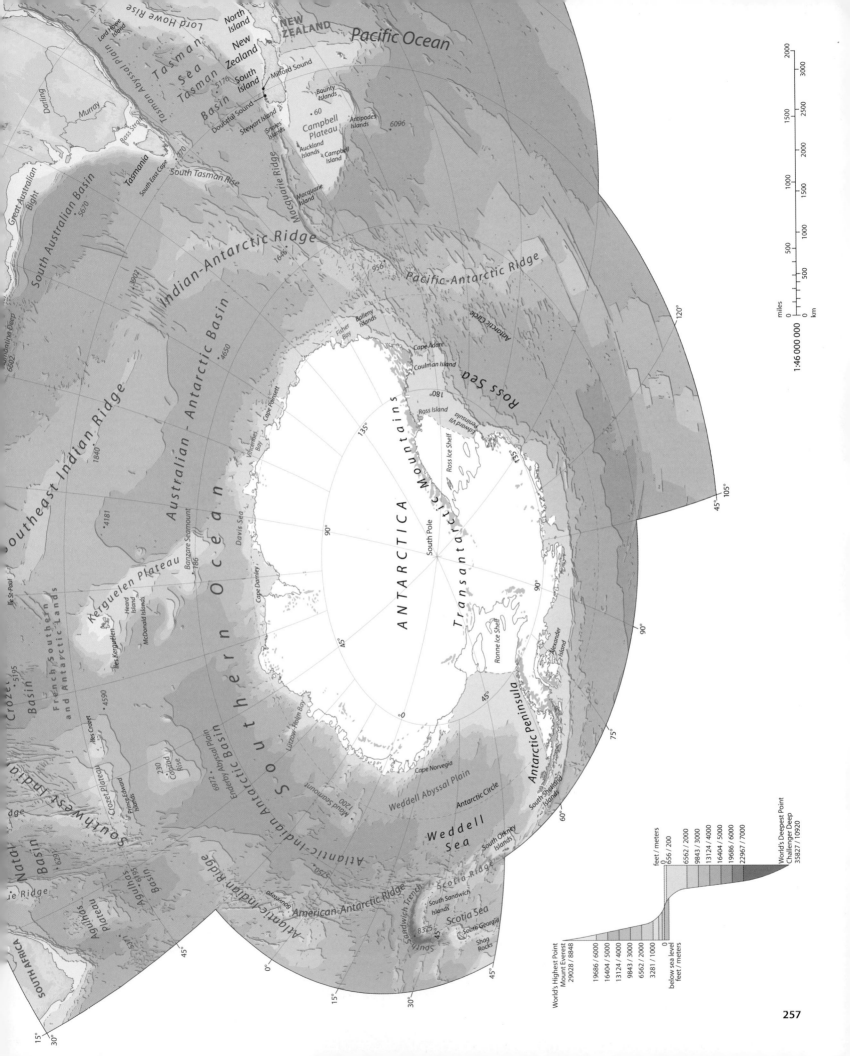

Lord Howe Rise
Lord Howe Island
Tasman Sea
North Island
NEW ZEALAND
New Zealand
Pacific Ocean
South Island
Milford Sound
Doubtful Sound
5176
Stewart Island
South Tasman Basin
Snares Islands
Bounty Islands
·60
Antipodes Islands
6096
Campbell Plateau
Auckland Islands
Campbell Island

Darling
Murray
Great Australian Bight
Tasman Abyssal Plain
Bass Strait
770
5570
South Australian Basin
Tasmania
South East Cape
South Tasman Rise
Macquarie Ridge
Macquarie Island

Agulhantina Deep
6602
Southeast Indian Ridge
Indian-Antarctic Ridge
3902
1649
956
Pacific-Antarctic Ridge
Antarctic Circle
180°

4650
Australian – Antarctic Basin
Fisher Bay
Balleny Islands
Cape Adare
Coulman Island
Ross Sea

1840
Southern Ocean
Vincennes Bay
Cape Forrett
Ross Island
Edward VII Peninsula
Ross Ice Shelf
135°

4181
135°
90°
ANTARCTICA
Transantarctic Mountains
Ile St-Paul
Kerguelen Plateau
Banzare Seamount
186
Davis Sea
Cape Darnley
South Pole
90°

French Southern and Antarctic Lands
Heard Island
McDonald Islands
Iles Kerguelen
45°
45°
0°
Ronne Ice Shelf
Alexander Island
90°

Crozet 5195
Crozet Basin
·4590
Iles Crozet
230
Copand Rise
Prince Edward Islands
Crozet Plateau
Enderby Abyssal Plain
Lützow-Holm Bay
45°
Antarctic Peninsula
75°

Natal Basin
629
SOUTH AFRICA
Agulhas Plateau
Southwest Indian Ridge
Ie Ridge
6972
Maud Seamount
1200
Cape Norvegia
Weddell Abyssal Plain
Antarctic Circle
South Shetland Islands
60°

Aguinhas Basin
Atlantic-Indian Ridge
5750
Atlantic-Indian Antarctic Ridge
Bouvetøya
American-Antarctic Ridge
Weddell Sea
South Orkney Islands
Scotia Ridge

South Sandwich Trench
8325
45°
South Sandwich Islands
Scotia Sea
South Georgia
Shag Rocks

15°
30°
0°
15°
30°
45°
45°
105°
120°
45°

1:46 000 000

miles
km
2000
3000
1500
2500
1000
1500
500
1000
500
0

257

feet / meters
656 / 200
6562 / 2000
9843 / 3000
13124 / 4000
16404 / 5000
19686 / 6000
22967 / 7000
World's Deepest Point
Challenger Deep
35827 / 10920

World's Highest Point
Mount Everest
29028 / 8848
19686 / 6000
16404 / 5000
13124 / 4000
9843 / 3000
6562 / 2000
3281 / 1000
below sea level
0
feet / meters

PACIFIC OCEAN

World's Highest Point
Mount Everest
29028 / 8848

19686 / 6000
16404 / 5000
13124 / 4000
9843 / 3000
6562 / 2000
3281 / 1000

feet / meters
0
656 / 200

below sea level
feet / meters

0
6562 / 2000
9843 / 3000
13124 / 4000
16404 / 5000
19686 / 6000
22967 / 7000

World's Deepest Point
Challenger Deep
35827 / 10920

miles
0   500   1000   1500   2000   2500
1:56 000 000
0   500   1000   1500   2000   2500   3000   3500   4000
km

Arctic Circle
Alaska
(U.S.A.)
Mackenzie
Hudson Bay
James Bay
CANADA
Cape Sable
Sable Island
Bay of Fundy
Gulf of Maine
Mid-Atlantic Ridge
ROCKY MOUNTAINS
Gulf of Alaska
Kodiak Island
1546
Alexander Archipélago
Queen Charlotte Islands
Vancouver Island
Columbia
Missouri
St. Lawrence
Elliot Lake
Georges Bank
5029
Great South Channel
Cape Cod
Gulf of Maine Bay
New England Seamounts
Stellwagen Bank
1092
Cornal Seamounts
Bermuda Rise
30°
Tufts Abyssal Plain
Mt. St. Helens
UNITED STATES OF AMERICA
Colorado
Susquehanna River
Chesapeake Bay
Nares Deep
6667
Tropic of Cancer
Cape Mendocino
2733
San Francisco
Monterey Bay
Rio Grande
Mississippi
New Orleans
Cape Hatteras
Cape Lookout
4556
Hatteras Abyssal Plain
Bermuda
Nares Abyssal Plain 5508
Sargasso Sea
Atlantic Ocean
15°
30°
Northeast Pacific
6217
Guadalupe
Golfo de California
Baja California
Gulf of Mexico
3504
Sigsbee Deep
Florida
Straits of Florida
Bahama Islands
Greater Antilles
Cuba
Milwaukee Deep
8605 Deep
Puerto Rico Trench
5523
Islands
hua'i
O'ahu
Maui
Hawai'i
Rancho Nuevo
MEXICO
Islas Revillagigedo
Isla Clarión
Isla Socorro
Palenque
Yucatan Channel
Tikal
GUATEMALA
Copán
EL SALVADOR
6662
Jamaica
Cayman Trench
7535
Hispaniola
Caribbean Sea
Venezuelan Basin
Lesser Antilles
Demerara Abyssal Plain
4923
idge
7022
Ile Clipperton
Nine North
Golfo de Tehuantepec
Tehuantepec Ridge
Middle America Trench
NICARAGUA
Guatemala Basin
COSTA RICA
Isthmus of Panama
PANAMA
Colombian Basin
Orinoco
Amazon Cone
East Pacific Basin
East Pacific Rise
Isla de Coco
Cocos Ridge
Isla de Malpelo
3901
COLOMBIA
abuaeran
Kiritimati
ands
Colon Ridge
Galápagos Islands
Carnegie Ridge
ECUADOR
Amazon
Equator
0°
Starbuck Island
Malden Island
Gallego Rise
Galapagos Rise
BRAZIL
Penrhyn Basin
Penrhyn
Vostok Island
Flint Island
Nuku Hiva
Iles Marquises
Hiva Oa
Caroline Island
French
Archipel des Tuamotu
6601
PERU
Peru Basin
Chincha Islands
Manuae
Iles Palliser
Raroia
1929
4385
Tiki Basin
Peru-Chile Trench
Nazca Ridge
Southwest Peru Ridge
Raiatea
Tahiti
Anaa
Hao
Polynesia
Tahitian Islands
Hervey Islands
Mangaia
rotonga
Maria
Iles du Duc
de Gloucester
Héréhérétué
Moruroa
Groupe
Actéon
Iles Gambier
5470
ANDES
Tubuai
Iles Australes
Raivavae
Pitcairn Island
Henderson Island
Ducie Island
Pitcairn Islands (U.K.)
Atacama Desert
Rapa
Isla Sala y Gómez
Isla de Pascua
1344
571
8170
30°
15°
Southwest
5420
Pacific Basin
Challenger Fracture Zone
Roggeveen Basin
Isla San Félix
Isla San Ambrosio
Chile Basin
Tropic of Capricorn
Archipiélago Juan Fernández
2743
Chile Rise
Parand
Santos Plateau
Pacific - Antarctic Ridge
4359
Mornington Abyssal Plain
Argentine Rise
30°
30°
Ocean
Southeast Pacific Basin
5230
Amundsen Abyssal Plain
Amundsen Ridges
Amundsen Sea
Peter I Island
Antarctic Circle
4325
Drake Passage
Cape de Hornos
Falkland Islands
Falkland Plateau
Argentine Abyssal Plain
6041
45°
ICA
135°

259

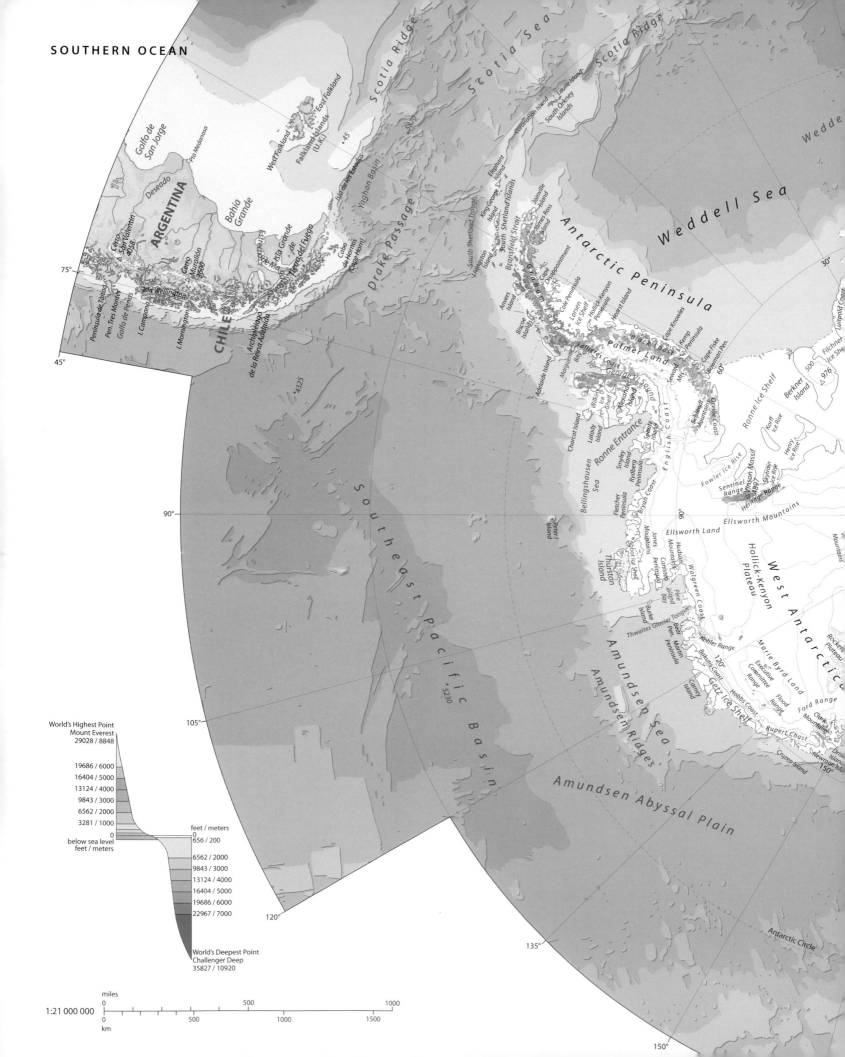

SOUTHERN OCEAN

Golfo de San Jorge

Pta Medanosa

ARGENTINA

Deseado

Bahía Grande

West Falkland
Falkland Islands (U.K.)
East Falkland

·45

Scotia Ridge

Scotia Sea

Scotia Ridge

Coronation Island
Laurie Island
South Orkney Islands

·2870

Wedde

Weddell Sea

Cerro San Valentín
4058
Cerro Murallón
3600

Isla Grande de Tierra del Fuego

Cabo de Hornos (Cape Horn)

Islas de los Estados

Estrecho de Ma

Yaghan Basin

Elephant Island
King George Island
Jorville Island
James Ross Island

South Shetland Islands

Bransfield Strait

Cape Disappointment

Antarctic Peninsula

30°

75°

CHILE

Isla Wellington
I. Campana
Pen. Tres Montes
Golfo de Penas
I. Morrington

Peninsula de Taitao
Archipélago de la Reina Adelaida

Drake Passage

4325

South Shetland Trough

Livingston Island
Anvers Island

Graham Land

Larsen Ice Shelf
Cole Peninsula
Hollick-Kenyon Peninsula
Hearst Island

Black Coast

Cape Knowles

Kemp Peninsula

Cape Fiske

Palmer Land

Seward Mts.
Bowman Pen.

60°

Filchner
Ice She

Biscoe Islands

Adelaide Island

Marguerite Bay

Fallieres Coast

George VI Sound

Wilkins Ice Shelf
Alexander Island

Latady Island
Charcot Island

Smyley Island

Rydberg Peninsula

Ronne Entrance

Sports Island

Behrendt Mountains

English Coast

Orville Coast

Ronne Ice Shelf

500

Berkner Island

976

Fowler Ice Rise

Korff Ice Rise

Henry Ice Rise

45°

Bellingshausen Sea

Fletcher Peninsula

Bryan Coast

Peter I Island

·06

Sentinel Range
Heritage Range

Vinson Massif
4897
Skytrain Ice Rise

90°

Southeast Pacific Basin

Ellsworth Land

Ellsworth Mountains

West Antarctic

Hollick-Kenyon Plateau

Thurston Island

Abbot Ice Shelf

Hudson Mountains

Jones Mountains

Canisteo Peninsula

Burke Island

Bear Island

Pine Island Bay

Walgreen Coast

Marie Byrd Land

Rockef
Plateau

·5230

Thwaites Glacier Tongue

Martin Peninsula

Kohler Range

Bakutis Coast

Executive Committee Range

Flood Range

Ford Range

105°

Amundsen Ridges

Amundsen Sea

Carney Island

Getz Ice Shelf

Hobbs Coast

Clark Mountains

Rupert Coast

Cruzen Island

Newman

World's Highest Point
Mount Everest
29028 / 8848

19686 / 6000
16404 / 5000
13124 / 4000
9843 / 3000
6562 / 2000
3281 / 1000
0
below sea level
feet / meters

feet / meters
0
656 / 200
6562 / 2000
9843 / 3000
13124 / 4000
16404 / 5000
19686 / 6000
22967 / 7000

Amundsen Abyssal Plain

120°

135°

Antarctic Circle

World's Deepest Point
Challenger Deep
35827 / 10920

miles
0   500   1000
1:21 000 000
0   500   1000   1500
km

150°

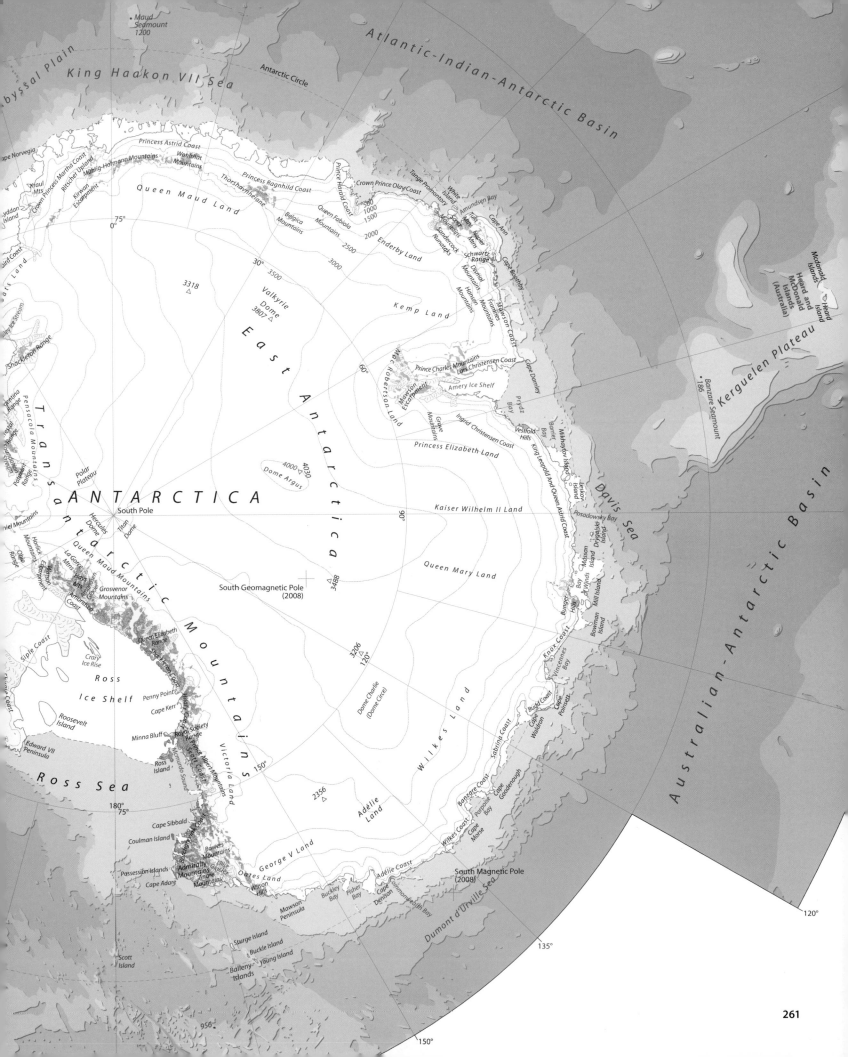

Maud Seamount
1200

Abyssal Plain

King Haakon VII Sea

Antarctic Circle

Cape Norvegia

Kraul Mts

Crown Princess Martha Coast

Ritscher Upland

Princess Astrid Coast

Mühlig-Hofmann Mountains

Wohlthat Mountains

Kirwan Escarpment

Yddan Island

Queen Maud Land

Thorshavnheiane

Princess Ragnhild Coast

Prince Harald Coast

Crown Prince Olav Coast

Tange Promontory

White Island

Amundsen Bay

McKaskins

Tula Mtns

Napier Mtns

Cape Ann

Belgica Mountains

Queen Fabiola Mountains

Enderby Land

Sandercock Nunataks

Schwartz Range

Cape Boothby

75°

0°

East Antarctica

Valkyrie Dome
3807

3318

30°

3500

3000

2500

2000

1500

1000

Kemp Land

Dismal Mountains

Hansen Mountains

Framnes Mountains

Mawson Coast

Prince Charles Mountains

Lars Christensen Coast

60°

Mawson Escarpment

Amery Ice Shelf

Cape Darnley

Kerguelen Plateau

McDonald Islands

Heard and McDonald Islands (Australia)

Heard Island

Banzare Seamount
186

Shackleton Range

Argentina Range

Pensacola Mountains

Forrestal Range

Neptune Range

Patuxent Range

Transantarctic

Grove Mountains

Ingrid Christensen Coast

Robertson Land

Princess Elizabeth Land

Prydz Bay

Vestfold Hills

Barrier Bay

King Leopold And Queen Astrid Coast

Mikhaylov Island

Davis Sea

Leskov Island

Drygalski Island

4000
4030

Dome Argus

Polar Plateau

ANTARCTICA

South Pole

90°

Kaiser Wilhelm II Land

Queen Mary Land

Posadowsky Bay

Masson Island

Mill Island

Bowman Island

Australian-Antarctic Basin

Daniel Mountains

Horlick Mountains

Ohio Range

Hercules Dome

Titan Dome

Queen Maud Mountains

La Gorce Mtns

Hays Mtns

Grosvenor Mountains

South Geomagnetic Pole (2008)

3488

Watson Escarpment

Amundsen Coast

Queen Elizabeth Range

Shackleton Coast

Crary Ice Rise

Ross Ice Shelf

Siple Coast

Roosevelt Island

Edward VII Peninsula

Ross Sea

Penny Point

Cape Kerr

Minna Bluff

Royal Society Range

Ross Island

Hillary Coast

Prince Albert Mountains

Scott Coast

Victoria Land

Mountains

120°

3206

Dome Charlie (Dome Circe)

Wilkes Land

Bunger Hills

Vincennes Bay

Knox Coast

Budd Coast

Cape Waldron

Sabrina Coast

Cape Poinsett

2356

Adélie Land

150°

Cape Sibbald

Coulman Island

Passession Islands

Cape Adare

Bowers Mountains

Admiralty Mountains

Anare Mountains

Little Glacier

Wilson Hills

Oates Land

George V Land

Mawson Peninsula

Buckley Bay

Fisher Bay

Cape Denison

Adélie Coast

Commonwealth Bay

South Magnetic Pole (2008)

Banzare Coast

Porpoise Bay

Cape Goodenough

Wilkes Coast

Cape Morse

180°

75°

Sturge Island

Buckle Island

Balleny Islands

Young Island

Scott Island

135°

Dumont d'Urville Sea

120°

956

150°

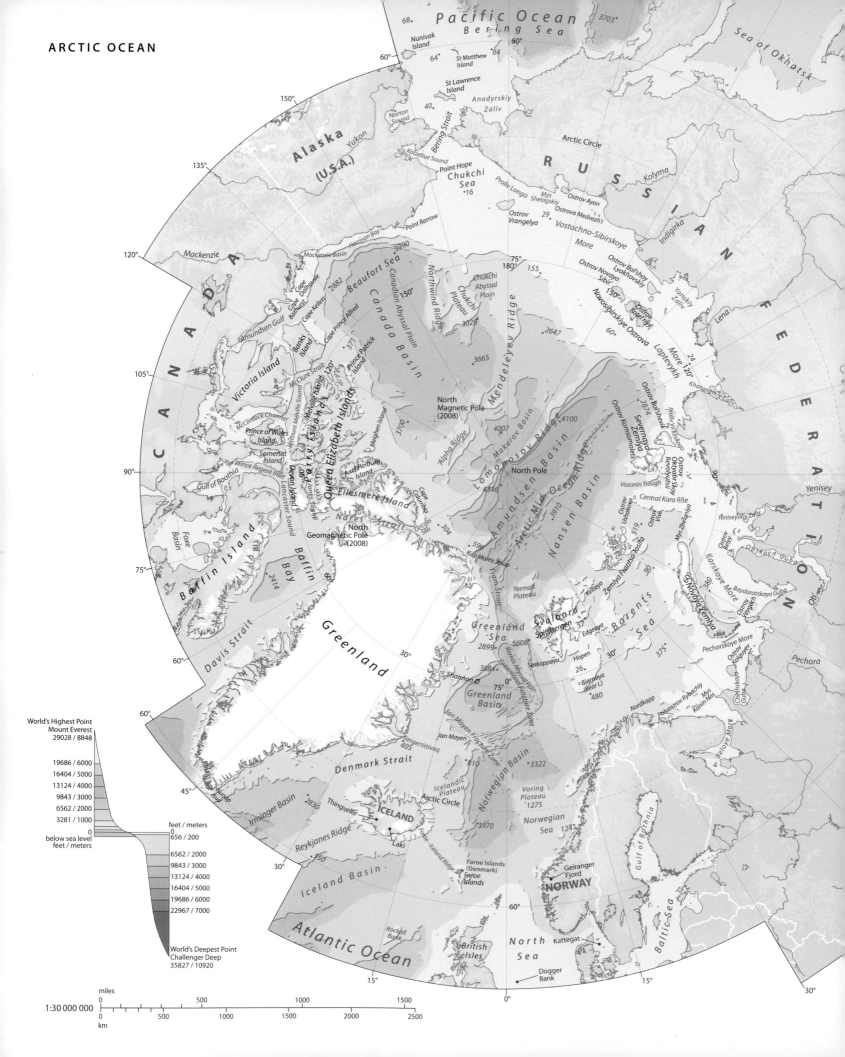

Supplementary Maps

Note: Names printed in red on the maps are discussed in the text.

● Fossil site

Arctic

SIBERIA

Arctic Circle

Beaufort
Sea

Somerset Island

Chukchi
Sea

Victoria
Island

Mt McKinley
6194

● Acasta River

Bering

Sea

Gulf
of Alaska

Salient
Mountain
● Burgess Shale
The Great Plains ●

N C

ROCKY

MOUNTAINS

Lake Superi

AME

Aleutian Islands

Aleutian Trench

A S I A

Gobi
Desert

Kuril Trench

Japan Trench

Northeast Pacific Basin

Lena

Lake Baikal

Amur

40°

60°

80°

Yellow

Meishan
Cliffs

Yangtze

● Doushantuo Formation
● Lake Dianchi

Tropic of Cancer

20°

South
China
Sea

Challenger Deep
10920

Mariana Trench

Hawai'ian Islands

Colorado

Missouri

El Capitan ●

G
of M

Mid-Pacific Mountains

P a c i f i c

0° Equator

Borneo

Sumatera

Java Trench

Java

7445

Sulawesi

Puncak Jaya ▲ New
5030   Guinea

O c e a n

Indian
Ocean

West Australian

Basin

Arafura
Sea

● Roper
Valley

Coral

Sea

Great Barrier Reef

Fiji Islands

East Pacific Rise

Chi

20°

Tropic of Capricorn

● Pilbara Craton

A U S T R A L I A

Great Dividing Range

Tonga Trench

Southwest

Pacific

Perth

Basin

Ediacara Hills ●

Darling

Tasman

Sea

New Zealand

Basin

40°

Southeast
Indian Ridge

South Australian Basin

Tasmania

Pacific-Antarctic Ridge

Southeast Pacif

Australian-Antarctic Basin

S o u t h e r n   O c e a n

Amundsen Sea

60°

Davis Sea

Antarctic Circle

Antarctic Mountains

Ross
Sea

80°

A N T

O c e a n

Ellesmere   Sirius Passet
Island

Greenland

Baffin   Greenland
Bay   Sea

Baffin   Barents
Island   Sea

Norwegian
Sea

Arctic Circle

Hudson   Labrador   Ural Mountains
Bay   Sea

Ob'

North   60°
British   Sea
RTH   Isles
Gunflint   Gros Morne   White Cliffs   Chalk Cliffs   E U R O P E   Volga   Ob'
Chert   Newfoundland   of Dover   of France
Jasper   Spaniard's Bay   Alps   El'brus   40°
Nob   Restigouche   Mistaken Point   Pesciara   Danube   Black Sea   5642   Caspian Sea   Kunlun Shan
RICA   Estuary   A S I A   HIMALAYA
Appalachian   Mediterranean Sea   Troodos   Tigris   Indus
Mountains   Atlas Mountains   Mountains   Ganges   Mt Everest

A t l a n t i c   The Gulf   8850

Nares   S A H A R A   Tropic of Cancer
Deep

Milwaukee Deep   Nile   Red Sea   Oman   20°
ico   8605   Bay
Greater Antilles   S A H E L   A r a b i a n   of
Caribbean Sea   Cape Verde   Sea   Bengal
            Basin   Niger   A F R I C A

Orinoco   Gulf of Guinea   Lake   Equator   0°
                            Congo   Congo   Victoria   I n d i a n
Amazon   Basin   Kilimanjaro
O c e a n   5895   Mid-Indian
SOUTH   Basin
Lake   O c e a n
Titicaca   Zambezi
AMERICA   Mozambique Channel
Parana   Namibia   Kalahari   Madagascar   20°
Cerro Aconcagua   Desert   Tropic of Capricorn
6959   Barberton
Peru-Chile Trench   Pinnacle Point

Rise   Cape   Crozet   40°
            Basin   Basin
Argentine Basin   Southeast
                      Agulhas   Indian Ridge
                      Basin

Scotia Sea

Drake Passage   Atlantic-Indian-Antarctic Basin   60°

Antarctic   Antarctic Circle
Peninsula
Vinson Massif   Weddell Sea
4897

A R C T I C A

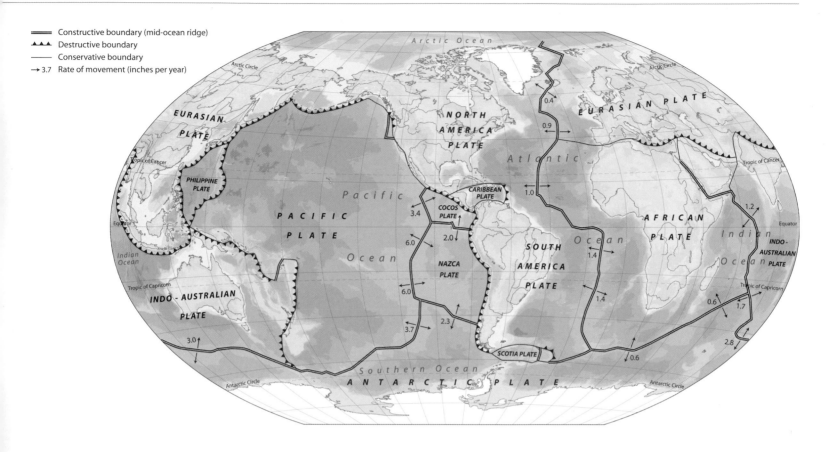

- ═══ Constructive boundary (mid-ocean ridge)
- ▲▲▲ Destructive boundary
- ── Conservative boundary
- → 3.7 Rate of movement (inches per year)

HYDROTHERMAL VENTS *Map 3*

- Vent fields confirmed by human or camera observation

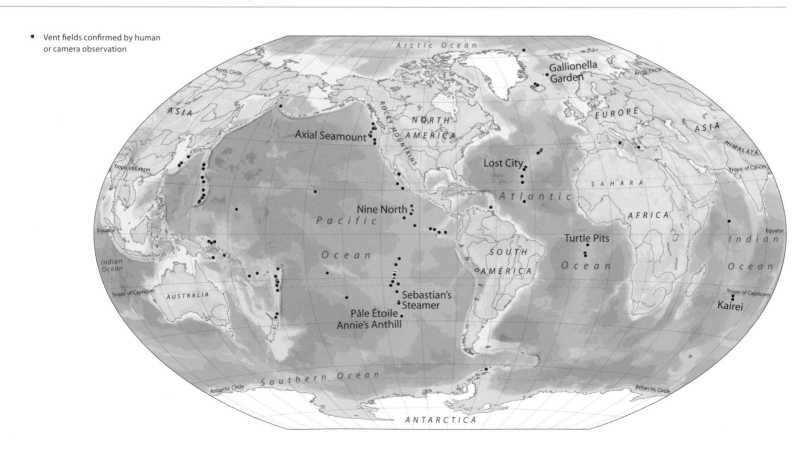

- Earthquake
  (greater than magnitude 7.5 on Richter scale)

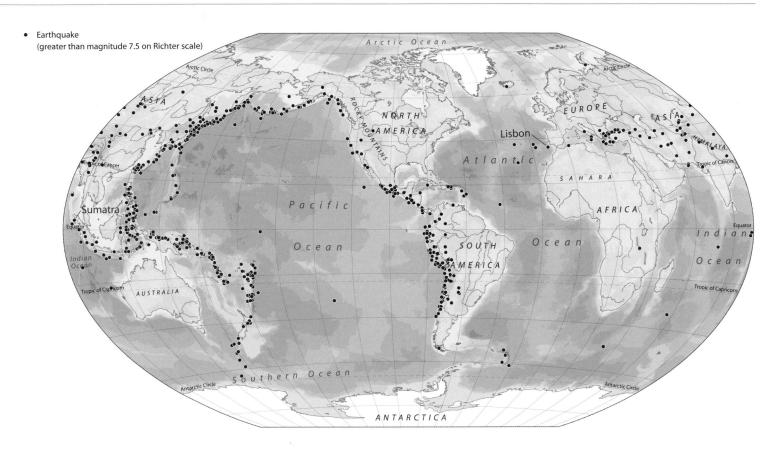

▲ Volcano

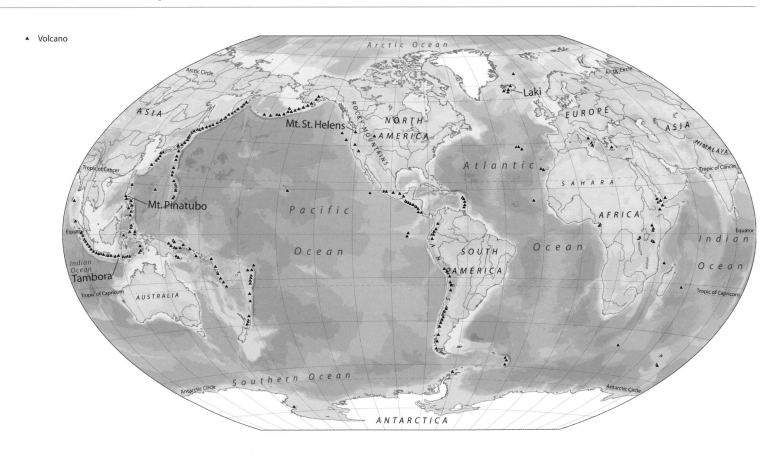

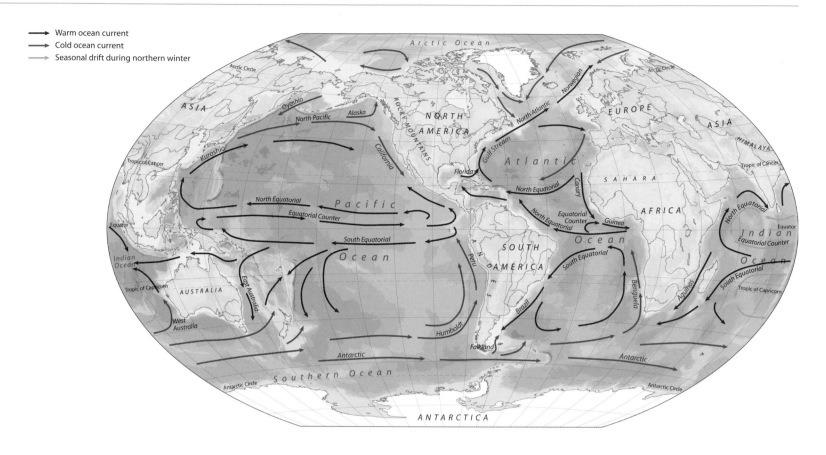

Warm ocean current
Cold ocean current
Seasonal drift during northern winter

OCEAN CONVEYOR BELT *Map 11*

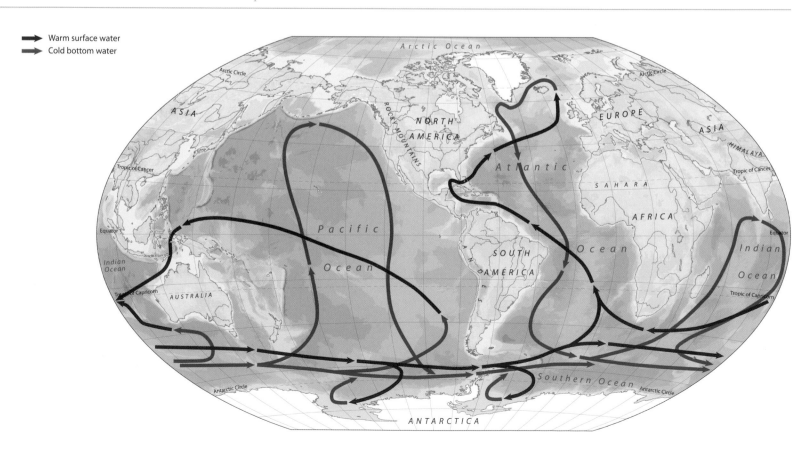

Warm surface water
Cold bottom water

Major port
Shipping route

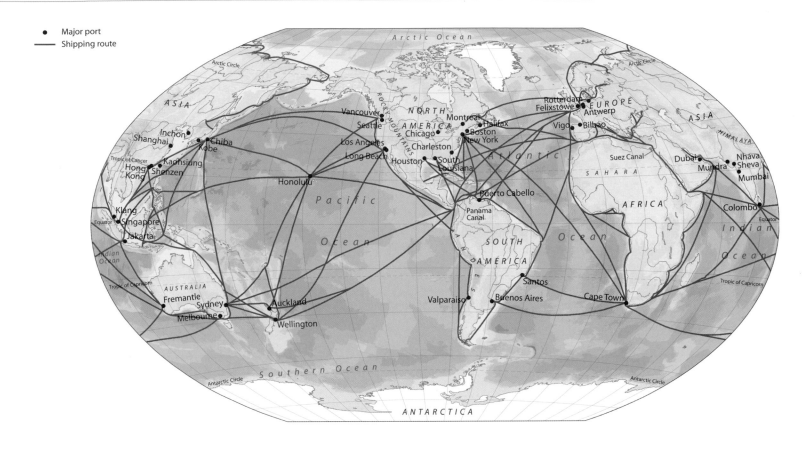

FISH STOCK EXPLOITATION *Map 13*

Exploitation status of marine fish stocks, analyzed by FAO area, and combined for a global picture

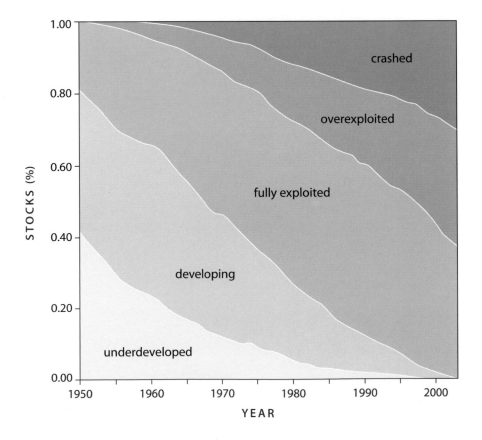

Metric tons per square kilometer

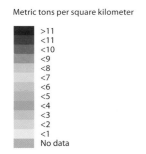

>11
<11
<10
<9
<8
<7
<6
<5
<4
<3
<2
<1
No data

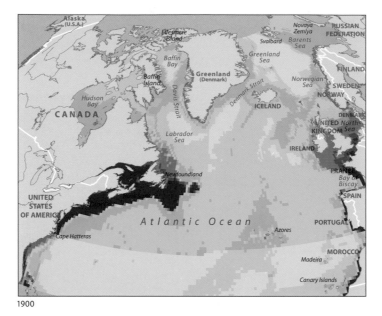

1900

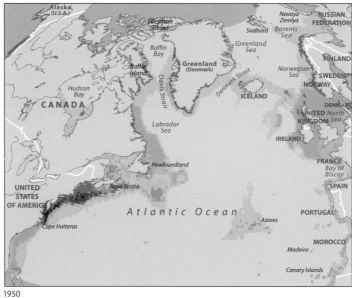

1950

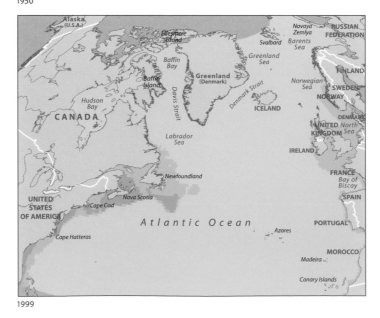

1999

Calving ground
Feeding ground
→ Migration route

CANADA

Gulf of St Lawrence

Newfoundland

Lake Huron

Lake Ontario

Summer/Fall

Bay of Fundy

Nova Scotia

Cape Sable

Winter/Spring

Cape Cod

Summer/Fall

Lake Erie

Long Island

Spring/Summer

UNITED STATES OF AMERICA

Delaware Bay

Chesapeake Bay

Atlantic Ocean

Cape Hatteras

Cape Fear

Cape Romain

Fall/Winter

Florida

LETHAL WHALE STRIKES OFF THE COAST OF NORTH AMERICA, 2002–2006 *Map 16*

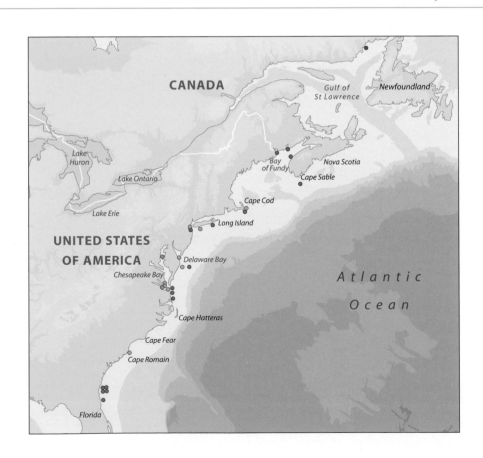

● Fin
● Humpback
● Minke
● Right
● Sei

CANADA

Gulf of St Lawrence

Newfoundland

Lake Huron

Lake Ontario

Bay of Fundy

Nova Scotia

Cape Sable

Cape Cod

Lake Erie

Long Island

UNITED STATES OF AMERICA

Delaware Bay

Chesapeake Bay

Atlantic Ocean

Cape Hatteras

Cape Fear

Cape Romain

Florida

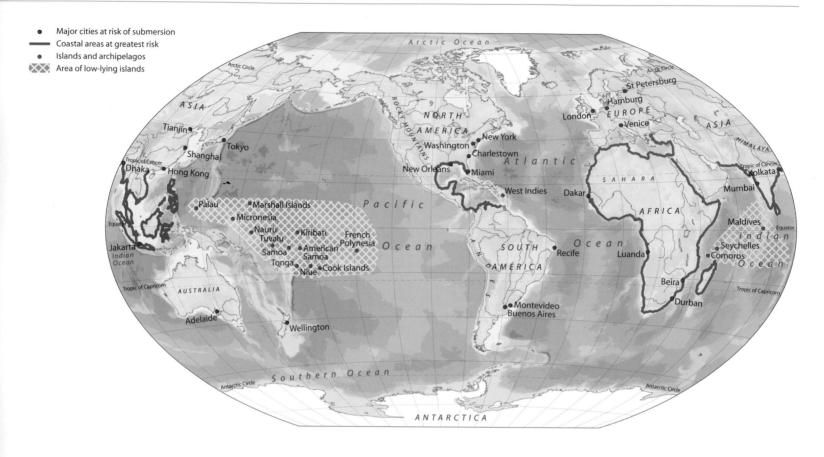

- Major cities at risk of submersion
- Coastal areas at greatest risk
- Islands and archipelagos
- Area of low-lying islands

ARCTIC SEA ICE CONCENTRATION   *Map 18*

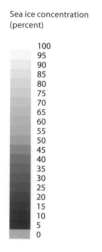

Sea ice concentration
(percent)

100
95
90
85
80
75
70
65
60
55
50
45
40
35
30
25
20
15
10
5
0

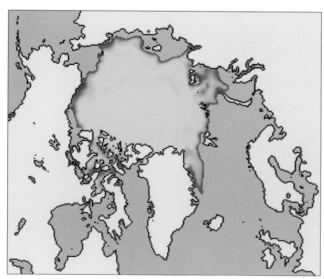

September 1980
Total area: 3.03 million sq miles/7.85 million sq km

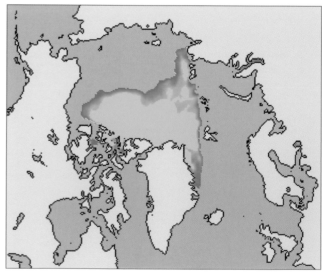

September 2007
Total area: 1.65 million sq miles/4.28 million sq km

Maps | Timeline Credits

Except where indicated in the text, all maps and diagrams are copyright Collins Bartholomew Ltd.

Plate tectonic and paleogeographic maps (pp. 74, 95, 122, and 144) by C. R. Scotese, © 2005, PALEOMAP Project. www.scotese.com.

DATA SOURCES

Fossil sites
p. 264: Daniel G. Cole, GIS Coordinator, Smithsonian Institution.

Hydrothermal vents
p. 266: Edward T. Baker, NOAA, Pacific Marine Environmental Laboratory.

Earthquakes
p. 267: United States Geological Survey, National Earthquakes Information Center, Denver.

Coral reefs
p. 268: *Reefs at Risk*. Washington, D.C.: World Resources Institute; Cambridge, U.K.: UNEP-WCMC.

Salt marshes
p. 268: Modified from Chapman, V. J. 1977. *Wetland Coastal Ecosystems*. New York: Elsevier Scientific.

Mangroves
p. 269: UNEP World Conservation Monitoring Centre/Global Polygon Version 3.0 Dataset.

Marine protected areas
p. 269: Wood, L. J. 2007. MPA Global: A database of the world's marine protected areas. Sea Around Us Project, UNEP-WCMC and WWF. www.mpaglobal.org.

Ocean conveyor belt
p. 270: Rick Lumpkin, NOAA, Atlantic Oceanographic and Meteorological Laboratory; Anand Gnanadesikan, NOAA, Geophysical Fluid Dynamics Laboratory.

Fish stock exploitation
p. 271: Sea Around Us Project, University of British Columbia Fisheries Centre, initiated and funded by the Pew Charitable Trusts, Philadelphia. www.seaaroundus.org.

Change in fish biomass
p. 272: Sea Around Us Project, University of British Columbia Fisheries Centre, initiated and funded by the Pew Charitable Trusts, Philadelphia. www.seaaroundus.org.

North Atlantic right whale feeding and calving grounds
p. 273: E. Paul Oberlander, Woods Hole Oceanographic Institution.

Lethal whale strikes
p. 273: Glass, A. H., T. V. N. Cole, M. Garron, R. L. Merrick, and R. M. Pace III. 2008. Mortality and serious injury determinations for baleen whale stocks along the United States eastern seaboard and adjacent Canadian Maritimes, 2002–2006. Northeast Fisheries Science Center Reference Document 08-04. www.nefsc.noaa.gov/publications/crd/crd0804/.

Arctic sea ice concentration
p. 274: National Snow and Ice Data Center, Boulder.

TIMELINE

Geologic dates
p. 277: National Museum of Natural History, www.paleobiology.si.edu/geotime/main/.

Geologic time clock
p. 277: Hannes Grobe, Alfred Wegener Institute for Polar and Marine Research, Bremerhaven 27515, Germany.

Aggregate anemone (Anthopleura *sp.*) *in waters off San Diego, California.*

GEOLOGIC TIMELINE

EON	ERA	PERIOD		EPOCH	MILLION YRS AGO	MAJOR EVENTS
PHANEROZOIC	CENOZOIC	Quaternary		Holocene	0	Rise of *Homo sapiens*
				Pleistocene	0.01	Ice Ages begin
		Tertiary	Neogene	Pliocene	1.8	Isthmus of Panama rises
				Miocene	5.3	Human ancestors in Africa
			Paleogene	Oligocene	23	Seals and sea lions appear
				Eocene	33.9	First whales India collides with Asia
				Paleocene	55.8	Beginnings of many modern mammal lineages
	MESOZOIC	Cretaceous			65.5	First flowering plants Extinction of dinosaurs
		Jurassic			146	Atlantic Ocean begins to form First birds
		Triassic			200	First dinosaurs First mammals
	PALEOZOIC	Permian			252	Formation of Pangaea Major mass extinction
		Carboniferous	Pennsylvanian		299	Abundant coal-forming swamps First reptiles
			Mississippian		318	Four-legged vertebrates begin to diversify
		Devonian			359	First four-legged vertebrates
		Silurian			416	First land plants
		Ordovician			444	First coral reefs
		Cambrian			488	Explosion of animal life
PRECAMBRIAN	PROTEROZOIC				542	First multicellular organisms Banded iron formations Earliest animals with shells
	ARCHEAN				2,500	Earliest fossil record of life Photosynthesizing organisms emerge
	HADEAN				4,000	Creation of the earth, its continents and ocean

4,600

MILLION YRS AGO

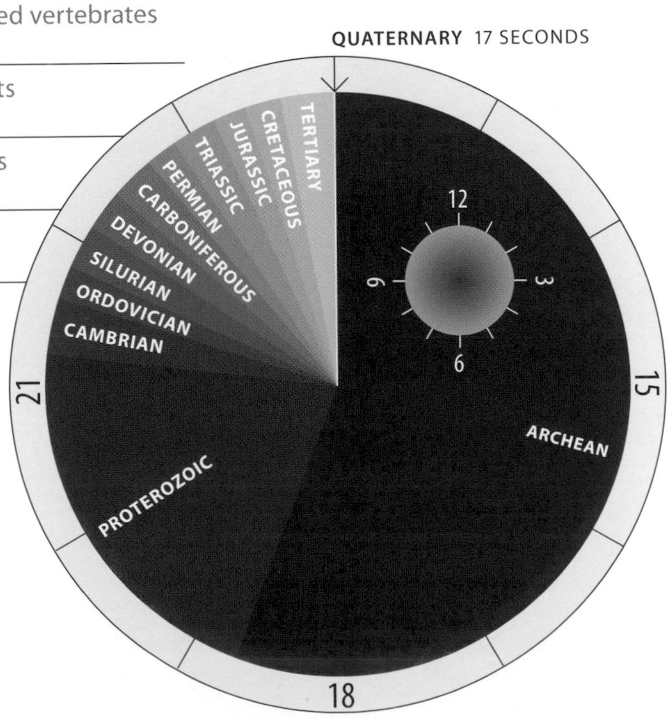

QUATERNARY 17 SECONDS

GEOLOGIC TIME MAPPED TO 24-HOUR CLOCK

Acknowledgments

Mantle of a giant clam (Tridacna gigas), 40 feet (12 m) deep, Papua New Guinea.

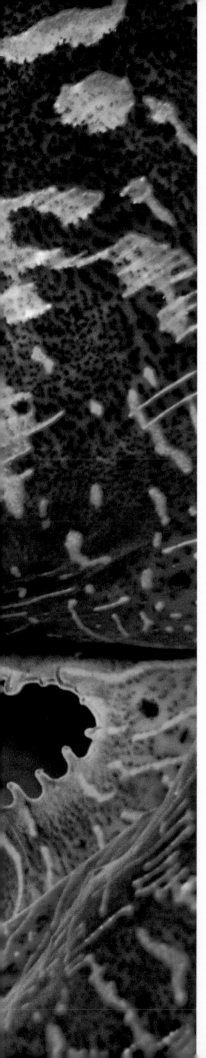

It is a great honor to have been asked by the Smithsonian's National Museum of Natural History to write *Smithsonian Ocean* and a privilege to work with its dedicated scientists. I am particularly grateful to Director Cristián Samper, whose commitment to the book's themes was there from the beginning.

The pivotal and prescient suggestions of Hans-Dieter Sues, the museum's associate director of research and collections, helped shape the book's structure. His perspicacity, wide-ranging knowledge, and thoughtful reviews of every chapter were invaluable. Doug Erwin from the Department of Paleobiology helped conceptualize and generously reviewed the "Bridging Past and Present" sections. Nancy Knowlton's scientific research and commitment to the sea have been inspirations long before she became the museum's Sant chair in marine science. Brian Huber, Carole Baldwin, Michael Vecchione, Jill Johnson, and Carolyn Margolis generously shared mission statements, reference material, and exhibit scripts, enabling the incorporation of exhibit themes into the book's narrative backbone. David Bohaska, William Fitzhugh, Igor Krupnik, James Mead, Ross Robertson, and Jean-Daniel Stanley reviewed text, and other Smithsonian scientists also shared their knowledge and resources: Bruce Collette, Adrienne Kaeppler, Ian Macintyre, David Pawson, Charles Potter, Robert Purdy, Victor Springer, and George Zug.

Scientists from outside the Smithsonian also gave their time, expertise, and resources, and generously reviewed chapters or sections of text: David Ainley, Richard Aronson, Ann Bucklin, Jennifer Clack, Lee Cooper, Peter deMenocal, Robert Diaz, Chuck Fisher, Alan Friedlander, Joel Garlich-Miller, Chris German, Michael Graham, Rosemary Grant, Ove Hoegh-Guldberg, Amy Knowlton, Matthew Lawrence, William Leavenworth, Richard Limeburner, Dede Marx, Marilyn Marx, Walt Meier, Randall Miller, Jonathan Overpeck, Daniel Pauly, Larry Peterson, Pamela Plotkin, James Price, Peter Pritchard, Luiz Rocha, Enric Sala, Stuart Sandin, Adam Schultz, Michel Segonzac, Degan Shu, Craig Smith, Julienne Stroeve, Curtis Suttle, Mark Uhen, Harvey Weiss, and Reg Wilson.

In addition, many other scientists answered questions and provided references: Abigail Allwood, Neil Banerjee, Tom Beatty, Jeff Bolster, Jochen Brocks, Roger Buick, Nicholas Butterfield, Ted Daeschler, Maarten de Wit, Steven D'Hondt, Greg Donovan, Jennifer Eigenbrode, Jason Flores, Michael Fogarty, Sonja Fordham, Margaret Fraiser, Patricia Gensel, Rachel Haymon, Hans Hofmann, Russ Hopcroft, Julie Huber, Deborah S. Kelley, Andy Knoll, Andrea Koschinsky, Joe Jones, David Lund, Enrique Macpherson, Bramley Murton, Sean Nee, Martin Nweeia, Jonathan Payne, Pete Peterson, Susannah Porter, Will Robbins, Minik Rosing, William Royer, Ray Schmitt, Mahmood Shivji, Mitchell Sogin, Hubert Staudigel, Peter Thomas, Eugene Turner, Robert C. Vrijenhoek, Wes Watters, and Shuhai Xiao.

Over the years, MIT has supported my work in many ways. A visiting scholar appointment in its Earth System Initiative made it possible for me to undertake the extensive research required for this book.

There are two whose work deeply informs mine. Science now demonstrates, again and again, how farsighted and encompassing was Rachel Carson's understanding of the sea, expressed more than fifty years ago in *The Sea Around Us*. Ransom Myers of Dalhousie University died in 2007. He contributed originally, and greatly, to our understanding of world fisheries, and we will miss him.

Many thanks to Brian Huber for suggesting and to Chip Clark for photographing fossils to feature in the book; to Mary Sears from the Ernst Mayr Library at Harvard's Museum of Comparative Zoology; to Michael Houck and Michael Delano from the Smithsonian's technical support team; to Megan Miller; to Carol Cumming, David White, Sarah Woods, Louisa Wood, and Dirk Zeller for beautiful maps; to editor Duke Johns for utmost care and steadfast endurance; to the unflappable Steve and Deborah Freligh, Bob Tope, and Gillian Ginter from Nature's Best Publishing, and Miriam Stein for gathering the beautiful photography; to designer Bill Anton for his boundless creativity and keen aesthetic sense; and to Caroline Newman, executive editor at Smithsonian Books, who brought together this committed and talented team. From her wise leadership came a book whose whole is far greater than the sum of its parts. I am grateful for the enthusiasm and dedication of agent Wendy Strothman and publisher HarperCollins.

Lastly, thank you to Robert Buchsbaum for help when time ran short, to Laila Goodman for critical reading, to Abby and Susannah for each solving a seemingly unsolvable problem, and to Dan, whose loyalty, energy, and support never wavers.

Sources

The following bibliography contains references and further details for the animals, ecosystems, and geologic events described in the text. Sources that are repeated are shortened after their first citation in full.

1. EDEN IN THE DEPTHS

Pulses in the Deep

Tolstoy, M., J. R. Cowen, E. T. Baker, et al. 2006. A sea-floor spreading event captured by seismometers. *Science* 314:1920–1922.

Tubeworm Barbecue and Life Reemerges from the Ruins

Delaney, J. R., D. S. Kelley, M. D. Lilley, et al. 1998. The quantum event of oceanic crustal accretion: Impacts of diking at mid-ocean ridges. *Science* 281:222–230.

Haymon, R. M., D. J. Fornari, K. L. Von Damm, et al. 1993. Volcanic eruption of the mid-ocean ridge along the East Pacific Rise crest at 9°45–52′N: Direct submersible observations of seafloor phenom-ena associated with an eruption event in April, 1991. *Earth and Planetary Science Letters* 119:85–101.

The Staff of Life

Fisher, C. R., and P. Girguis. 2007. A proteomic snapshot of life at a vent. *Science* 315:198–199.

Flores, J. F., C. R. Fisher, S. L. Carney, et al. Sulfide binding is mediated by zinc ions discovered in the crystal structure of a hydrothermal vent tubeworm hemoglobin. *Proceedings of the National Academy of Sciences* 102: 2713–2718.

Lutz, R. A., T. M. Shank, D. J. Fornari, et al. 1994. Rapid growth at deep-sea vents. *Nature* 371: 663–664.

Nussbaumer, A. D., C. R. Fisher, and M. Bright. 2006. Horizontal endosymbiont transmission in hydrothermal vent tubeworms. *Nature* 441:345–348.

Shank, T. M., D. J. Fornari, K. L. Von Damm, et al. 1998. Temporal and spatial patterns of biological community development at nascent deep-sea hydrothermal vents (9°50′N, East Pacific Rise). *Deep-Sea Research* II 45:465–515.

A Faint Light

Beatty, J. T., J. Overmann, M. T. Lince, et al. 2005. An obligately photosynthetic bacterial anaerobe from a deep-sea hydrothermal vent. *Proceedings of the National Academy of Sciences* 102:9306–9310.

Gebruk, A.V., E. C. Southward, H. Kennedy, et al. 2000. Food sources, behaviour, and distribution of hydrothermal vent shrimps at the Mid-Atlantic Ridge. *Journal of the Marine Biological Association of the United Kingdom* 80: 485–499.

Pond, D. W., A. Gebruk, E. C. Southward, et al. 2000. Unusual fatty acid composition of storage lipids in the bresilioid shrimp *Rimicaris exoculata* couples the photic zone with MAR hydrothermal vent sites. *Marine Ecology Progress Series* 198:171–179.

White, S. N., A. D. Chave, G. T. Reynolds, et al. 2002. Ambient light emission from hydrothermal vents on the Mid-Atlantic Ridge. *Geophysical Research Letters* 29, 744, doi:10.1029/2002GL014977.

A New Sea Is Born

Alemseged, Z., F. Spoor, W. H. Kimbel, et al. 2006. A juvenile early hominin skeleton from Dikika, Ethiopia. *Nature* 443: 296–301.

de Heinzelin, J., J. D. Clark, T. White, et al. 1999. Environment and behavior of 2.5-million-year-old Bouri hominids. *Science* 284:625–629.

Haile-Selassie, Y. 2001. Late Miocene hominids from the Middle Awash, Ethiopia. *Nature* 412:178–181.

Wright, T. J., C. Ebinger, J. Biggs, et al. 2006. Magma-maintained rift segmentation at continental rupture in the 2005 Afar dyking episode. *Nature* 442:291–294.

Beyond the Volcanoes of the Ridge

Bergquist, D. C., F. M. Williams, and C. R. Fisher. 2000. Longevity record for deep-sea invertebrate. *Nature* 403: 499–500.

Boetius, A. 2005. Lost City life. *Science* 307:1420–1421.

Cordes, E. E., M. A. Arthur, K. Shea, et al. 2005. Modeling the mutualistic interactions between tubeworms and microbial consortia. *PLoS Biology* 3:497–506.

Früh-Green, G. L., D. S. Kelley, S. M. Bernasconi, et al. 2003. 30,000 years of hydrothermal activity at the Lost City vent field. *Science* 301:495–498.

Kelley, D. S., J. A. Karson, G. L. Früh-Green, et al. 2005. A serpentinite-hosted ecosystem: The Lost City hydro-thermal field. *Science* 307:1428–1434.

Continual Revelation

Desbruyères, D., M. Segonzac, and M. Bright, eds. 2006. *Handbook of Deep-Sea Hydrothermal Vent Fauna.* 2nd ed. *Denisia* 18. Linz: Biologiezentrum der Oberöster-reichischen Landesmuseen. Supplement and corrigendum, December 2006, www.noc.soton.ac.uk/chess/hand-book.php.

Huber, J. A., D. B. M. Welch, H. G. Morrison, et al. 2007. Microbial population structures in the deep marine biosphere. *Science* 318:97–100.

Koschinsky, A., A. Billings, C. Devey, et al. 2006. Discovery of new hydrothermal vents on the southern Mid-Atlantic Ridge (4°S–10°S) during cruise M68/1. *InterRidge News* 15:9–15.

MacPherson, E., W. Jones, and M. Segonzac. 2006. A new squat lobster family of Galatheoidea (Crustacea, Decapoda, Anomura) from the hydrothermal vents of the Pacific-Antarctic Ridge. *Zoostema* 27:709–723.

Reed, C. 2006. Boiling points. *Nature* 439:905–907.

Warén, A., S. Bengtson, S. K. Goffredi, et al. 2003. A hot-vent gastropod with iron sulfide dermal sclerites. *Science* 302:1007.

Lifelines

Little, C. T. S., and R. C. Vrijenhoek. 2003. Are hydrothermal vent animals living fossils? *Trends in Ecology and Evolution* 18:582–588.

Murton, B. J., E. T. Baker, C. M. Sands, et al. 2006. Detection of an unusually large hydrothermal event plume above the slow-spreading Carlsberg Ridge: NW Indian Ocean. *Geophysical Research Letters* 33, LI0608, doi:10.1029/2006GL026048.

Van Dover, C. L., C. R. German, K. G. Speer, et al. 2002. Evolution and biogeography of deep-sea vent and seep invertebrates. *Science* 295:1253–1257.

Left and pages 291, 292: Kelp, Channel Islands National Marine Sanctuary and National Park, California.

Bridging Past and Present:
FIRST LAND, FIRST SEA

A Hellish Start
Fortey, R. 1999. *Life: A Natural History of the First Four Billion Years of Life on Earth*. New York: Vintage.

Grotzinger, J., T. H. Jordan, F. Press, and R. Siever. 2007. *Understanding Earth*. 5th ed. New York: W. H. Freeman.

Hazen, R. M. 2005. Genesis: *The Scientific Quest for Life's Origin*. Washington, D.C.: Joseph Henry Press.

Knoll, A. H. 2003. *Life on a Young Planet: The First Three Billion Years of Evolution on Earth*. Princeton: Princeton University Press.

The Land beside the Sea
Banerjee, N. R., H. Furnes, K. Muehlenbachs, et al. 2006. Preservation of ~3.4–3.5 Ga microbial biomarkers in pillow lavas and hyaloclastites from the Barberton Greenstone Belt, South Africa. *Earth and Planetary Science Letters* 241:707–722.

Furnes, H., M. de Wit, H. Staudigel, et al. 2007. A vestige of earth's oldest ophiolite. *Science* 315:1704–1707.

Grotzinger et al., *Understanding Earth*.

Another Hellish Start
Fortey, *Life*.

Hazen, *Genesis*.

Kashefi, K., and D. R. Lovley. 2003. Extending the upper temperature limit for life. *Science* 301:934.

Knoll, *Life on a Young Planet*.

Reysenbach, A.-L., and E. Shock. 2002. Merging genomes with geochemistry in hydrothermal ecosystems. *Science* 296:1077–1082.

Vestiges
Allwood, A. C., M. R. Walter, B. S. Kamber, et al. 2006. Stromatolite reef from the Early Archaean era of Australia. *Nature* 441:714–718.

Banerjee et al., Preservation of ~3.4–3.5 Ga microbial biomarkers.

Delaney et al., Quantum event of oceanic crustal accretion.

Grotzinger et al., *Understanding Earth*.

Haymon et al., Volcanic eruption of the mid-ocean ridge.

Hazen, *Genesis*.

Philippot, R., M. V. Zuilen, K. Lepot, et al. 2007. Early Archaean microorganisms preferred elemental sulfur, not sulfate. *Science* 317:1534–1537.

Rasmussen, B. 2000. Filamentous microfossils in a 3235-million-year-old volcanogenic massive sulphide deposit. *Nature* 405:676–679.

Endurance
Baker, B. J., G. W. Tyson, R. I. Webb, et al. Lineages of acidophilic Archaea revealed by community genomic analysis. *Science* 314:1933–1936.

Bohannon, J. 2005. Microbe may push photosynthesis into deep water. *Science* 308:1855.

Huber et al., Microbial population structures.

Nisbet, E. 2000. The realms of Archaean life. *Nature* 405:625–626.

2. BUILDING THE BASIN

Darwin, C. 1909. *Voyage of the Beagle*. New York: P. F. Collier and Son. Chap. 15, "Passage of the Cordillera," March 20, 1835.

The Floor of the Sea:
From the Mountains to the Plains
Maury, M. F. 1855. *The Physical Geography of the Sea*. New York: Harper and Brothers.

Nybakken, J. W., and M. D. Bertness. 2005. *Marine Biology: An Ecological Approach*. 6th ed. San Francisco: Pearson Benjamin Cummings.

Earth's Most Voluminous Habitat:
Cold, Dark, Deep
Colette, B. B., and G. Klein-MacPhee. 2002. *Bigelow and Schroeder's Fishes of the Gulf of Maine*. 3rd ed. Washington, D.C.: Smithsonian Institution Press.

Gillibrand, E. J. V., P. Bagley, A. Jamieson, et al. 2007. Deep sea benthic bioluminescence at artificial food falls, 1,000–4,800 m depth, in the Porcupine Seabight and Abyssal Plain, North East Atlantic Ocean. *Marine Biology* 150:1053–1060.

Nybakken and Bertness, *Marine Biology*.

Sumich, J. L., and J. F. Morrissey. 2004. *Introduction to the Biology of Marine Life*. 8th ed. Sudbury, Mass: Jones and Bartlett.

Lighting the Deep Sea
Lupp, C., and E. G. Ruby. 2005. *Vibrio fischeri* uses two quorum-sensing systems for the regulation of early and late colonization factors. *Journal of Bacteriology* 187:3620–3629.

McFall-Ngai, M. J., and E. G. Ruby. 2000. Developmental biology in marine invertebrate symbioses. *Current Opinion in Microbiology* 3:603–607.

Robinson, B. H., K. R. Reisenbichler, J. C. Hunt, et al. 2003. Light production by the arm tips of the deep-sea cephalopod *Vampyroteuthis infernalis*. *Biological Bulletin* 205:102–109.

Wonders Unfolding
Biddle, J. F., J. S. Lipp, M. A. Lever, et al. 2006. Heterotrophic Archaea dominate sedimentary subsurface ecosystems off Peru. *Proceedings of the National Academy of Sciences* 103:3846–3851.

Cherel, Y., and G. Duhamel. 2004. Antarctic jaws: Cephalopod prey of sharks in Kerguelen waters. *Deep-Sea Research* I 51:17–31.

Douglas, R. H., J. C. Partridge, K. Dulai, et al. 1998. Dragon fish see using chlorophyll. *Nature* 393: 422–423.

Haddock, S. H. D., C. W. Dunn, P. R. Pugh, et al. 2005. Bioluminescent and red-fluorescent lures in a deep-sea siphonophore. *Science* 309:263.

Jørgensen, B. B., and S. D'Hondt. 2006. A starving majority deep beneath the seafloor. *Science* 314:932–933.

Seibel, B. A., B. H. Robison, and S. H. D. Haddock. 2005. Post-spawning egg care by a squid. *Nature* 438:929.

From the Plains of the Abyss to the
Dry Land of Continents
Grotzinger et al., *Understanding Earth*.

Gutscher, M.-A. 2004. What caused the Great Lisbon Earthquake? *Science* 305:1247–1248.

Han, S.-C., C. K. Shum, M. Bevis, et al. 2006. Crustal dilation observed by GRACE after the 2004 Sumatra-Andaman earthquake. *Science* 313:658–659.

Kerr, R. A. 2006. Stealth tsunami surprises Indonesian coastal residents. *Science* 313:742–743.

Stone, R. 2006. The day the land tipped over. *Science* 314:406–409.

Landscape Redesigned by
a Closing Ocean Basin
Furnes et al., Earth's oldest ophiolite.

Grotzinger et al., *Understanding Earth*.

Bridging Past and Present:
BREATHING

Earth's Young Sea: Iron-Rich
Knoll, *Life on a Young Planet*.

Reed, Boiling points.

A Whiff of Fresh Air
Allen, J. F., and W. Martin. 2007. Out of thin air. *Nature* 445:610–612.

Des Marais, D. J. 2000. Evolution: When did photosynthesis emerge on Earth? *Science* 289:1703–1705.

Falkowski, P. G. 2006. Tracing oxygen's imprint on Earth's metabolic evolution. *Science* 311:1724–1725.

Margulis, L., and M. F. Dolan. 2002. *Early Life: Evolution on the PreCambrian Earth*. 2nd ed. Sudbury, Mass.: Jones and Bartlett.

Those That Imbued the Atmosphere
with Oxygen
Brocks, J. J., G. A. Logan, R. Buick, et al. 1999. Archean molecular fossils and the early rise of eukaryotes. *Science* 285:1033–1036.

Butterfield, N. 2007 Macroevolution and macroecology through deep time. *Palaeontology* 50:41–55.

Coleman, M. L., M. B. Sullivan, A. C. Martiny, et al. 2006. Genomic islands and the ecology and evolution of *Prochlorococcus*. *Science* 311:1768–1770.

Dutkiewicz, A., H. Volk, S. C. George, et al. 2006. Biomarkers from Huronian oil-bearing fluid inclusions: An uncontaminated record of life before the Great Oxidation Event. *Geology* 34:437–440.

Eigenbrode, J. L., and K. H. Freeman. 2006. Late Archean rise of aerobic microbial ecosystems. *Proceedings of the National Academy of Sciences* 103:15759–15764.

Fortey, *Life*.

Fuhrman, J. 2003. Genome sequences from the sea. *Nature* 424:1001–1002.

Knoll, *Life on a Young Planet*.

Pennisi, E. 2006. How a marine bacterium adapts to multiple environments. *Science* 311:1697.

Sullivan, M. B., J. B. Waterbury, and S. W. Chisholm. 2003. Cyanophages infecting the oceanic cyanobacterium *Prochlorococcus*. *Nature* 424:1047–1051.

Summons, R. E., L. L. Jahnke, J. M. Hope, et al. 1999. 2-Methylhopanoids as biomarkers for cyanobacterial oxygenic photosynthesis. *Nature* 400:554–557.

A Partnership
Kump, L. R., and M. E. Barley. 2007. Increased subaerial volcanism and the rise of atmospheric oxygen 2.5 billion years ago. *Nature* 448:1033–1036.

Lyons, T. L. 2007 Oxygen's rise reduced. *Nature* 448: 1005–1006.

Margulis and Dolan, *Early Life*.

Margulis, L., and D. Sagan. 2002. *Acquiring Genomes: A Theory of the Origins of Species*. New York: Basic Books.

Rosing, M. T., D. K. Bird, N. H. Sleep, et al. 2006. The rise of continents—An essay on the geologic consequences of photosynthesis. *Palaeogeography, Palaeoclimatology, Palaeoecology* 232:99–113.

The Sea Rusted
U.S. Geological Survey. 1999. The mineral industry of Michigan. In *Minerals Yearbook*, vol. 2. *Area Reports: Domestic*. Washington, D.C.: U.S. Department of the Interior, U.S. Geological Survey.

Intimation of Life to Come
Brocks et al., Archean molecular fossils.

Butterfield, N. J. 2000. *Bangiomorpha pubescens* n. gen., n. sp.: Implications for the evolution of sex, multicellularity, and the Mesoproterozoic/Neoproterozoic radiation of eukaryotes. *Paleobiology* 26:386–404.

Fortey, *Life*.

Knoll, *Life on a Young Planet*.

Earth's Barren Middle Ages
Anbar, A. D., and A. H. Knoll. 2002. Proterozoic ocean chemistry and evolution: A bioinorganic bridge? *Science* 297:1137–1142.

Arnold, G. L., A. D. Anbar, J. Barling, et al. 2004. Molybdenum isotope evidence for widespread anoxia in mid-Proterozoic oceans. *Science* 304:87-90.

Brocks, J. J., G. D. Love, R. E. Summons, et al. 2005. Biomarker evidence for green and purple sulphur bacteria in a stratified Palaeoproterozoic sea. *Nature* 437:866–870.

Kerr, R. A. 2002. Could poor nutrition have held life back? *Science* 297:1104–1105.

Knoll, *Life on a Young Planet*.

3. CIRCLE OF WATER

Dickson, R. R., and J. Brown. 1994. The production of North Atlantic Deep Water: Sources, rates, and pathways. *Journal of Geophysical Research 99* (C6): 12319–12342.

Macrander, A., U. Send, and R. H. Kaese. 2003. New direct overflow observations at the Denmark Strait sill. Abstract no. 12596. Abstracts from the EGS–AGU–EUG Joint Assembly in Nice, France, April 6–11.

Ubiquitous Sea
Sverdrup, K. A., A. C. Duxbury, and A. B. Duxbury. 2005. *An Introduction to the World's Oceans*. 8th ed. Boston: McGraw-Hill.

The Changing Face of Earth's Continents and Sea
Fortey, R. 2004. *Earth: An Intimate History*. New York: Alfred A. Knopf.

Scotese, C. 2000. "Pangea Ultima" will form 250 million years in the future. PALEOMAP Project. www.scotese. com/future2.htm.

Torsvik, T. H. 2003. The Rodinia jigsaw puzzle. *Science* 300:1379–1381.

Rhythms in the Deep
Garrison, T. 2006. *Essentials of Oceanography*. 4th ed. Belmont, Calif.: Thomson.

Open University Course Team. 2001. *Ocean Circulation*. 2nd ed. Oxford: Butterworth Heinemann.

Schmitz, W. J., Jr., and M. S. McCartney. 1993. On the North Atlantic circulation. *Reviews of Geophysics* 31:29–49.

Sumich and Morrissey, *Biology of Marine Life*.

Tracking the Currents
Lawrence, M. 2004. Lost and found: The search for the *Portland*. *Sea History* 107:19–21.

Marx, D. 2004. Forbidden to sail: The steamship *Portland*, 1890–1898. *Sea History* 107:16–18.

Open University Course Team, *Ocean Circulation*.

Along the Surface Drift Passengers Great and Small
Collerson, K. D., and M. I. Weisler. 2007. Stone adze compositions and the extent of ancient Polynesian voyaging and trade. *Science* 317:1907–1911.

Energy Information Administration. 2006. World energy and economic outlook. In *International Energy Outlook*. Report No. DOE/ EIA-0484. Washington, D.C.: U.S. Department of Energy.

Finney, Ben. 2007. Tracking Polynesian seafarers. *Science* 317:1873–1874.

Garrison, *Essentials of Oceanography*.

Grotzinger et al., *Understanding Earth*.

Kirk, R., and R. D. Daugherty. 1974. *Hunters of the Whale*. New York: William Morrow.

Lund, D. C., J. Lynch-Stieglitz, and W. B. Curry. 2006. Gulf Stream density structure and transport during the past millennium. *Nature* 444:601–604.

Quimby, G. 1985. Japanese wrecks, iron tools, and prehistoric Indians of the Northwest Coast. *Arctic Anthropology* 22:7–15.

Sumich and Morrissey, *Biology of Marine Life*.

A Drop of Water
Open University Course Team, *Ocean Circulation*.

Sumich and Morrissey, *Biology of Marine Life*.

Connections: Opening and Closing Seas, Flowing Currents
Bowen, B. W., A. Muss, L. A. Rocha, et al. 2006. Shallow mtDNA coalescence in Atlantic pygmy angelfishes (genus *Centropyge*) indicates a recent invasion from the Indian Ocean. *Journal of Heredity* 97:1–12.

Knowlton, N., L. A. Weigt, L. A. Solorzano, et al. 1993. Divergence in proteins, mitochondrial DNA, and reproductive compatibility across the Isthmus of Panama. *Science* 260:1629–1632.

National Geographic Society. 1995. *National Geographic Atlas of the World*. 6th ed. Washington, D.C.: National Geographic Society, 131.

Rocha, L. A., D. R. Robertson, C. R. Rocha, et al. 2005. Recent invasion of the tropical Atlantic by an Indo-Pacific coral reef fish. *Molecular Biology* 14:3921–3928.

Sverdrup, Duxbury, and Duxbury, *World's Oceans*.

Deluge
Goetz, D., and S. G. Morley. 1954. Part 1, Chap. 1. In *Popul Vuh: The Book of the People*. From Adrián Recino's translation from Quiché into Spanish. Los Angeles: Plantin Press. Available online at www. sacred-texts.com/nam/maya/pvgm/index.htm.

Open University Course Team, *Ocean Circulation*.

Bridging Past and Present: EXPLOSION OF LIFE

An Enigma
Narbonne, G. M. 2004. Modular construction of early Ediacaran complex life forms. *Science* 305:1141–1144.

———. 2005. The Ediacara biota: Neoproterozoic origin of animals and their ecosystems. *Annual Review of Earth and Planetary Sciences* 33:421–422.

Narbonne, G. M., and J. G. Gehling. 2003. Life after snowball: The oldest complex Ediacaran fossils. *Geology* 31:27–30.

Glimmers of a World to Come
Grotzinger, J. P., W. A. Watters, and A. H. Knoll. 2000. Calcified metazoans in the thrombolite-stromatolite reefs of the terminal Proterozoic Nama Group, Namibia. *Paleobiology* 26:334–359.

Hofmann, H. J., and E. W. Mountjoy. 2001. *Namacalathus-Cloudina* assemblage in Neoproterozoic Miette Group (Byng Formation), British Columbia: Canada's oldest shelly fossils. *Geology* 29:1091–1094.

Knoll, *Life on a Young Planet*.

Knoll, A. H., E. J. Javaux, D. Hewitt, et al. 2006. Eukaryotic organisms in Proterozoic oceans. *Philosophical Transactions of the Royal Society B*: 361:1023–1038.

Porter, S. M., R. Meisterfeld, and A. H. Knoll. 2003. Vase-shaped microfossils from the neoproterozoic Chuar Group, Grand Canyon: A classification guided by modern testate amoebae. *Journal of Paleontology* 77:409–429.

Xiao, S., J. W. Hagadorn, C. Zhou, et al. 2007. Rare helical spheroidal fossils from the Doushantuo Lagerstätte: Ediacaran animal embryos come of age? *Geology* 35:115–118.

Yin, L., M. Zhu, A. H. Knoll, X. Yuan, et al. 2007. Doushantuo embryos preserved inside diapause egg cysts. *Nature* 446:661–663.

Air and Nourishment from the Sea
Canfield, D. E., S. W. Poultron, and G. M. Narbonne. 2007. Late-Neoproterozoic deep-ocean oxygenation and the rise of animal life. *Science* 315:92–95.

Fike, D. A., J. P. Grotzinger, L. M. Pratt, et al. 2006. Oxidation of the Ediacaran ocean. *Nature* 444:744–747.

Kerr, R. A. 2006. A shot of oxygen to unleash the evolution of animals. *Science* 314:1529.

Squire, R. J., I. H. Campbell, C. M. Allen, et al. 2006. Did the Transgondwanan Supermountain trigger the explosive radiation of animals on earth? *Earth and Planetary Science Letters* 250:116–133.

The Blossoming of Animal Life
Chen, J., L. Ramskold, and G. Zhou. 1994. Evidence for monophyly and arthropod affinity of Cambrian giant predators. *Science* 264:1304–1308.

Gould, S. J. 1989. *Wonderful Life: The Burgess Shale and the Nature of History*. New York: W. W. Norton.

Hu, S. 2005. Taphonomy and palaeoecology of the Early Cambrian Chengjiang biota from Eastern Yunnan, China. *Berliner Paläobiologische Abhandlungen* 7:1–197.

Knoll, *Life on a Young Planet*.

Morris, S. C. 1999. *The Crucible of Creation: The Burgess Shale and the Rise of Animals*. Oxford: Oxford University Press.

Xian-guang, H., R. J. Aldridge, J. Bergström, et al. 2004. *The Cambrian Fossils of Chengjiang, China: The Flowering of Early Animal Life*. Malden, Mass.: Blackwell.

Lands Divided and Joined, Fossils Map Drifting Continents
Fortey, R. 2000. *Trilobite! Eyewitness to Evolution*. New York: Alfred A. Knopf.

A Sea Both Strange and Familiar
Fortey, *Life*.

Hu, Early Cambrian Chengjiang biota.

Nedin, C. 1999. *Anomalocaris* predation on non-mineralized and mineralized trilobites. *Geology* 27:987–990.

Vannier, J., and J. Chen. 2005. Early Cambrian food chain: New evidence from fossil aggregates in the Maotianshan Shale biota, SW China. *Palaios* 20:3–26.

4. A GREEN SEA

The Power of Phytoplankton
Falkowski, P. G., M. E. Katz, A. H. Knoll, et al. 2004. The evolution of modern eukaryotic phytoplankton. *Science* 305:354–360.

Hays, G. C., A. J. Richardson, and C. Robinson. 2005. Climate change and marine plankton. *Trends in Ecology and Evolution* 20:337–344.

Knoll, A. H., R. E. Summons, J. R. Waldbauer, et al. 2007. The geological succession of primary producers in the oceans. In *The Evolution of Primary Producers in the Sea*, P. G. Falkowski and A. H. Knoll, eds., 7-37. Burlington: Elsevier.

Nybakken and Bertness, *Marine Biology*.

Sumich and Morrissey, *Biology of Marine Life*.

The Biology of the Very Small
Angly, F. E., B. Felts, M. Breitbart, et al. 2006. The marine viromes of four oceanic regions. *PLoS Biology* 4:2121-2131.

Azam, F. 1998. Microbial control of oceanic carbon flux: The plot thickens. *Science* 280:694–696.

Fuhrman, J. A. 1999. Marine viruses and their biogeochemical and ecological effects. *Nature* 399:541–548.

Miller, S. D., S. H. D. Haddock, C. D. Elvidge, et al. 2005. Detection of a bioluminescent milky sea from space. *Proceedings of the National Academy of Sciences* 102:14181–14184.

Nealson, K. H., and J. W. Hastings. 2006. Quorum sensing on a global scale: Massive numbers of bioluminescent bacteria make milky seas. *Applied and Environmental Microbiology* 72: 2295–2297.

Nee, S. 2004. More than meets the eye. *Nature* 429: 804–805.

Sogin, M. L., H. G. Morrison, J. A. Huber, et al. 2006. Microbial diversity in the deep sea and the underexplored "rare biosphere." *Proceedings of the National Academy of Sciences* 103:12115-12120.

Suttle, C. A. 2005. Viruses in the sea. *Nature* 437: 356–361.

Grazers
Kunze, E., J. F. Dower, I. Beveridge, et al. 2006. Observations of biologically generated turbulence in a coastal inlet. *Science* 313:1768–1770.

Nybakken and Bertness, *Marine Biology*.

Osborn, K. J., G. W. Rouse, S. K. Goffredi, et al. 2007. Description and relationships of *Chaetopterus pugaporcinus*, an unusual pelagic polychaete (Annelida, Chaetopteridae). *Biological Bulletin* 212:40–54.

Vannier, J., M. Steiner, E. Renvoise, et al. 2007. Early Cambrian origin of modern food webs: Evidence from predator arrow worms. *Proceedings of the Royal Society B* 274:627–633.

Mostly Water
Census of Marine Life. 2006. Extreme life, marine style, highlights 2006 ocean census. www.coml.org/medres/2006cn.htm.

Robison, B. H., K. R. Reisenbichler, and R. E. Sherlock. 2005. Giant larvacean houses: Rapid carbon transport to the deep sea floor. *Science* 308:1609–1611.

Woods Hole Oceanographic Institution. 2007. A photo gallery of animals from the Census of Marine Life zooplankton cruise. www.whoi.edu/page.do?pid=11458&tid=201&cid=10272&ct=362#.

A Sea within a Sea
Columbus, C. 1992. *The Log of Christopher Columbus*. Trans. R. H. Fuson. Camden, Maine: International Marine.

Nybakken and Bertness, *Marine Biology*.

Sumich and Morrissey, *Biology of Marine Life*.

Sverdrup, Duxbury, and Duxbury, *World's Oceans*.

Teal, J., and M. Teal. 1975. *The Sargasso Sea*. Boston: Atlantic Monthly Press.

Fertile Waters
Jahncke, J., D. M. Checkley Jr., and G. L. Hunt Jr. 2004. Trends in carbon flux to seabirds in the Peruvian upwelling system: Effects of wind and fisheries on population regulation. *Fisheries Oceanography* 13:208–223.

Sumich and Morrissey, *Biology of Marine Life*.

Bridging Past and Present:
ARRIVAL OF FISHES

Origins
Clack, J. A. 2002. *Gaining Ground: The Origin and Evolution of Tetrapods*. Bloomington: Indiana University Press.

Long, J. A. 1995. *The Rise of Fishes: 500 Million Years of Evolution*. Baltimore: Johns Hopkins University Press.

Maisey, J. G. 1996. *Discovering Fossil Fishes*. New York: Henry Holt.

Schultze, H.-P., and R. Cloutier, eds. 1996. *Devonian Fishes and Plants of Miguasha, Quebec, Canada*. Munich: Dr. Friedrich Pfeil.

Shu, D.-G., S. C. Morris, J. Han, et al. 2003. Head and backbone of the Early Cambrian vertebrate *Haikouichthys*. *Nature* 42:526–529.

Wilson, R. A., E. T. Burden, R. Bertrand, et al. 2004. Stratigraphy and tectono-sedimentary evolution of the Late Ordovician to Middle Devonian Gaspé Belt in northern New Brunswick: Evidence from the Restigouche area. *Canadian Journal of Earth Sciences* 41:527–551.

Jaws
Anderson, P. S. L., and M. W. Westneat. 2007. Feeding mechanics and bite force modeling of the skull of *Dunkleosteus terrelli*, an ancient apex predator. *Biology Letters* 3:76–79.

Clack, *Gaining Ground*.

Fortey, *Life*.

Long, *Rise of Fishes.*

Maisey, *Fossil Fishes.*

Miller, R. F., R. Cloutier, and S. Turner. 2003. The oldest articulated chondrichthyan from the Early Devonian period. *Nature* 425:501–504.

Miller, R. F. 2007. *Pterygotus anglicus* Agassiz (Chelicerata: Eurypterida) from Atholville, Lower Devonian Campbellton Formation, New Brunswick, Canada. *Palaeontology* 50:981–999.

Wilson et al., Stratigraphy and tectono-sedimentary evolution.

Bones
Clack, *Gaining Ground.*

Long, *Rise of Fishes.*

Maisey, *Fossil Fishes.*

Landfall
Ahlberg, P. E., and J. A. Clack. 2006. A firm step from water to land. *Nature* 440:747–749.

Clack, *Gaining Ground.*

Clack, J. A. 2005. The emergence of tetrapods. *Palaeogeography, Palaeoclimatology, Palaeoecology* 232:167–189.

Daeschler, E. B., N. H. Shubin, and F. A. Jenkins Jr. 2006. A Devonian tetrapod-like fish and the evolution of the tetrapod body plan. *Nature* 440:757–763.

Davis, M. C., R. D. Dahn, and N. H. Shubin. 2007. An autopodial-like pattern of Hox expression in the fins of a basal actinopterygian fish. *Nature* 447:473–476.

Fortey, *Life.*

Hotton, C. L., F. M. Hueber, D. H. Griffing, et al. 2001. Early terrestrial plant environments: An example from the Emsian of Gaspé, Canada. In *Plants Invade the Land: Evolutionary and Environmental Perspectives*, P. G. Gensel and D. Edwards, eds., 179–203. New York: Columbia University Press.

Knoll, *Life on a Young Planet.*

Long, J. A., G. C. Young, T. Holland, et al. 2006. An exceptional Devonian fish from Australia sheds light on tetrapod origins. *Nature* 444:199–202.

Wellman, C. H., P. L. Osterloff, and U. Mohiuddin. 2003. Fragments of the earliest land plants. *Nature* 425:282–285.

5. CLIMATE FROM THE SEA

Grotzinger et al., *Understanding Earth.*

El Niño: A Natural Rhythm in the Sea
Glantz, M. H. 2001. *Impacts of El Niño and La Niña on Climate and Society.* 2nd ed. Cambridge: Cambridge University Press.

Grant, B. R., and P. R. Grant. 2003. What Darwin's finches can teach us about the evolutionary origin and regulation of biodiversity. *Bioscience* 53:965–975.

Grant, P. R., and B. R. Grant. 2006. Evolution of character displacement in Darwin's finches. *Science* 313:224–226.

McPhaden, M. J., S. E. Zebiak, and M. H. Glantz. 2006. ENSO as an integrating concept in earth science. *Science* 314:1740–1745.

Yates, T. L., J. N. Mills, C. A. Parmenter, et al. 2002. The ecology and evolutionary history of an emergent disease: Hantavirus pulmonary syndrome. *Bioscience* 52:989–998.

The Rise of Agriculture
Colman, S. M. 2007. Conventional wisdom and climate history. *Proceedings of the National Academy of Sciences* 104:6500–6501.

Staubwasser, M., and H. Weiss. 2006. Holocene climate and cultural evolution in late prehistoric–early historic West Asia. *Quaternary Research* 66:372–387.

Weiss, H. 2000. Beyond the Younger Dryas. In *Environmental Disaster and the Archaeology of Human Response*, G. Bawden and R. M. Reycraft, eds., 75–98. Albuquerque: Maxwell Museum of Anthropology, University of New Mexico.

Weiss, H., and R. S. Bradley. 2001. What drives societal collapse? *Science* 291:609–610.

A Closing Sea Leaves Its Fingerprint on Earth's Climate
Gleckler, P. J., T. M. L. Wigley, B. D. Santer, et al. 2006. Volcanoes and climate: Krakatoa's signature persists in the ocean. *Nature* 439:675.

Grotzinger et al., *Understanding Earth.*

Kerr, R. A. 2006. Pollute the planet for climate's sake? *Science* 314:401–402.

The Collapse of the Maya
deMenocal, P. B. 2001. Cultural responses to climate change during the Late Holocene. *Science* 292:667–673.

Haug, G. H., D. Günther, L. C. Peterson, et al. 2003. Climate and the collapse of Maya civilization. *Science* 299:1731-1735.

Peterson, L. C., and G. H. Haug. 2005. Climate and the collapse of Maya civilization. *American Scientist* 93:322–329.

Feast and Famine in the Fertile Crescent
Cullen, H. M., and P. B. deMenocal. 2000. North Atlantic influence on Tigris–Euphrates streamflow. *International Journal of Climatology* 20:853–863.

Cullen, H. M., P. B. deMenocal, S. Hemming, et al. 2000. Climate change and the collapse of the Akkadian empire: Evidence from the deep sea. *Geology* 28:379–382.

deMenocal, Cultural responses to climate change.

Staubwasser and Weiss, Holocene climate.

The Greening, Then Drying of the Desert
Held, I. M., T. L. Delworth, J. Lu, et al. 2005. Simulation of Sahel drought in the 20th and 21st centuries. *Proceedings of the National Academy of Sciences* 102:17891–17896.

Kuper, R., and S. Kröpelin. 2006. Climate-controlled Holocene occupation in the Sahara: Motor of Africa's evolution. *Science* 313:803–807.

United Nations Environment Programme. 2007. Sudan: Post-conflict environmental assessment. Nairobi: United Nations Environment Programme.

Weldeab, S., R. R. Schneider, M. Kölling, et al. 2005. Holocene African droughts relate to eastern equatorial Atlantic cooling. *Geology* 33:981–984.

Taking Earth's Temperature
Falkowski, P. G. 2002. The ocean's invisible forest. *Scientific American* 287:54–62.

Falkowski, P., R. J. Scholes, E. Boyle, et al. 2000. The global carbon cycle: A test of our knowledge of earth as a system. *Science* 290:291–296.

Mayhew, P. J., G. B. Jenkins, and T. G. Benton. 2008. A long-term association between global temperature and biodiversity, origination and extinction in the fossil record. *Proceedings of the Royal Society B* 275:47–53.

Royer, D. L., R. A. Berner, and J. Park. 2007. Climate sensitivity constrained by CO_2 concentrations over the past 420 million years. *Nature* 446:530–532.

Bridging Past and Present: MASS EXTINCTION

Butterfield, Macroevolution.

Strange and Wondrous Sea
Davis, A. D., T. M. Weatherby, D. K. Hartline, et al. 1999. Myelin-like sheaths in copepod axons. *Nature* 398:571.

Erwin, D. 2006. *Extinction: How Life on Earth Nearly Ended 250 Million Years Ago.* Princeton: Princeton University Press.

Fortey, *Life.*

Snuffed Out
Butterfield, Macroevolution.

Erwin, *Extinction.*

Hallam, A. 2004. *Catastrophes and Lesser Calamities: The Causes of Mass Extinctions.* Oxford: Oxford University Press.

A Disaster of Earthly Origin
Erwin, *Extinction.*

Fraiser, M. L., and D. J. Bottjer. 2007. Elevated atmospheric CO_2 and the delayed biotic recovery from the end-Permian mass extinction. *Palaeogeography, Palaeoclimatology, Palaeoecology* 252:164–175.

Grice, K., C. Cao, G. D. Love, et al. 2005. Photic zone euxinia during the Permian-Triassic superanoxic event. *Science* 307:706–709.

Hallam, *Catastrophes.*

Knoll, A. H., R. K. Bambach, J. L. Payne, et al. 2007. Paleophysiology and end-Permian mass extinction. *Earth and Planetary Science Letters* 256:295–313.

Kump, L. R., A. Pavlov, and M. A. Arthur. 2005. Massive release of hydrogen sulfide to the surface ocean and atmosphere during intervals of oceanic anoxia. *Geology* 33:397–400.

Payne, J. L., D. J. Lehrmann, D. Follett, et al. 2007. Erosional truncation of upper-most Permian shallow-marine carbonates and implications for Permian-Triassic boundary events. *Geological Society of America Bulletin* 119:771–784.

Stone, R. 2004. Iceland's doomsday scenario? *Science* 306:1278–1281.

From Few, Many
Butterfield, Macroevolution.

Erwin, *Extinction*.

Fraiser, M. L., and D. J. Bottjer. 2007. When bivalves took over the world. *Paleobiology* 33:397–413.

Hallam, *Catastrophes*.

Peng, Y., G. R. Shi, Y. Gao, et al. 2007. How and why did the Lingulidae (Brachiopoda) not only survive the end-Permian mass extinction but also thrive in its aftermath? *Palaeogeography, Palaeoclimatology, Palaeoecology* 252:118–131.

Twitchett, R. J. 2007. The Lilliput effect in the aftermath of the end-Permian extinction event. *Palaeogeography, Palaeoclimatology, Palaeoecology* 252:132–144.

Legacies
Hallam, *Catastrophes*.

Simmons, S. F., and K. L. Brown. 2006. Gold in magmatic hydrothermal solutions and the rapid formation of a giant ore deposit. *Science* 314:288–291.

Extinction: Loss and Opportunity
Mayhew, Jenkins, and Benton. Global temperature and biodiversity.

Wagner, P. J., M. A. Kosnik, and S. Lidgard. 2006. Abundance distributions imply elevated complexity of post-Paleozoic marine ecosystems. *Science* 314:1289–1292.

THE ANTHROPOCENE

Bobe, R., and A. K. Behrensmeyer. 2004. The expansion of grassland ecosystems in Africa in relation to mammalian evolution and the origin of the genus *Homo*. *Palaeogeography, Palaeoclimatology, Palaeoecology* 207:399–420.

Cane, M. A., and P. Molnar. 2001. Closing of the Indonesian seaway as a precursor to east African aridification around 3–4 million years ago. *Nature* 411:157–162.

Chao, B. F. 1995. Anthropogenic impact on global geodynamics due to reservoir water impoundment. *Geophysical Research Letters* 22:3529–3532.

Crutzen, P. J. 2002. Geology of mankind. *Nature* 415:23.

deMenocal, P. B. 2004. African climate change and faunal evolution during the Pliocene-Pleistocene. *Earth and Planetary Science Letters* 220:3–24.

Federov, A. V., P. S. Dekens, M. McCarthy, et al. 2006. The Pliocene paradox (mechanisms for a permanent El Niño). *Science* 312:1485–1489.

Finlayson, C., F. G. Pacheco, J. Rodríguez-Vidal, et al. 2006. Late survival of Neanderthals at the southernmost extreme of Europe. *Nature* 443:850–853.

Halpern, B. S., S. Walbridge, K. A. Selkoe, et al. 2008. A global map of human impact on marine ecosystems. *Science* 319:948–952.

Haug, G. H., and R. Tiedemann. 1998. Effect of the formation of the Isthmus of Panama on Atlantic Ocean thermohaline circulation. *Nature* 393:673–676.

Kareiva, P., S. Watts, R. McDonald, et al. 2007. Domesticated nature: Shaping landscapes and ecosystems for human welfare. *Science* 316:1866–1869.

Krause, J., L. Orlando, D. Serre, et al. 2007. Neanderthals in central Asia and Siberia. *Nature* 449:902–904.

Mellars, P., B. Gravina, and C. Bronk Ramsey. 2007. Confirmation of Neanderthal/modern human interstratification at the Chatelperronian type-site. *Proceedings of the National Academy of Sciences* 104:3657–3662.

Potts, R. 1996. *Humanity's Descent: The Consequences of Ecological Instability*. New York: William Morrow.

Prideaux, G. J., R. G. Roberts, D. Megirian, et al. 2007. Mammalian responses to Pleistocene climate change in southeastern Australia. *Geology* 35:33–36.

Sarmiento, E., G. J. Sawyer, and R. Milner. 2007. *The Last Human: A Guide to Twenty-two Species of Extinct Humans*. New Haven: Yale University Press.

Sepulchre, P., G. Ramstein, F. Fluteau, et al. 2006. Tectonic uplift and eastern Africa aridification. *Science* 313:1419–1423.

Spoor, F., M. G. Leakey, P. N. Gathogo, et al. 2007. Implications of new early *Homo* fossils from Ileret, east of Lake Turkana, Kenya. *Nature* 448:688–691.

United Nations Environment Programme. 2007. *Global Environment Outlook (GEO) 4: Environment for Development*. Malta: Progress Press.

Wilson, E. O. 2002. *The Future of Life*. New York: Alfred A. Knopf.

——. 2006. *The Creation: An Appeal to Save Life on Earth*. New York: W. W. Norton.

Zalasiewicz, J., M. Williams, A. Smith, et al. 2008. Are we now living in the Anthropocene? *GSA Today* 18:4–8.

6. THE LONG MIGRATION

Bowen, B. W., R. A. Abreu-Grobois, G. H. Balazas, et al. 1995. Trans-Pacific migrations of the loggerhead turtle (*Caretta caretta*) demonstrated with mitochondrial DNA markers. *Proceedings of the National Academy of Sciences* 92:3731–3734.

Highsmith, R. C., K. O. Coyle, B. A. Bluhm, et al. 2007. Gray whales in the Bering and Chukchi seas. In *Whales, Whaling, and Ocean Ecosystems*, J. A. Estes, D. P. Demaster, D. F. Doak, et al., eds., 303–314. Berkeley: University of California Press.

Rasmussen, K., D. M. Palacios, J. Calambokidis, et al. 2007. Southern Hemisphere humpback whales wintering off Central America: Insights from water temperature into the longest mammalian migration. *Biology Letters* 3:302–305.

Back to the Sea
de Muizon, C. 2001. Walking with whales. *Nature* 413:259–260.

Gingerich, P. D., M. ul Haq, I. S. Zalmout, et al. 2001. Origin of whales from early artiodactyls: Hands and feet of Eocene Protocetidae from Pakistan. *Science* 293:2239–2242.

Hirayama, R. 1998. Oldest known sea turtle. *Nature* 392:705–708.

Lohmann, K. J., C. M. F. Lohmann, L. M. Ehrhart, et al. 2004. Geomagnetic map used in sea-turtle navigation. *Nature* 428:909–910.

Marino, L., R. C. Connor, R. E. Fordyce, et. al. 2007. Cetaceans have complex brains for complex cognition. *PLoS Biology* 5:966–972.

Rose, K. D. 2001. The ancestry of whales. *Science* 293:2216–2217.

Suzuki, R., J. R. Buck, and P. L. Tyack. 2006. Information entropy of humpback whale songs. *Journal of the Acoustical Society of America* 119:1849–1866.

Thewissen, J. G. M., E. M. Williams, L. J. Roe, et al. 2001. Skeletons of terrestrial cetaceans and the relationship of whales to artiodactyls. *Nature* 413:277–281.

Whitehead, H. 2007. Sperm whales in ocean ecosystems. In *Whales, Whaling, and Ocean Ecosystems*, Estes et al., eds., 324–334.

A Sea Filled with Turtles
Carr, A. F. 1984 [1967]. *So Excellent a Fishe: A Natural History of Sea Turtles*. New York: Charles Scribner's Sons.

Jackson, J. B. C. 1997. Reefs since Columbus. *Coral Reefs* 16, Suppl:S23–S32.

——. 2007. When ecological pyramids were upside down. In *Whales, Whaling, and Ocean Ecosystems*, Estes et al., eds., 27–37.

National Marine Fisheries Service, Office of Protected Resources, and U.S. Fish and Wildlife Service, Southwest Region. 2007. Kemp's ridley sea turtle (*Lepidochelys kempii*), 5-year review: Summary and evaluation. Silver Spring, Md.: National Marine Fisheries Service; Albuquerque: U.S. Fish and Wildlife Service.

Pritchard, P. C. H. 2007. *Arribadas* I have known. In *Biology and Conservation of Ridley Sea Turtles*, P. T. Plotkin, ed., 7–21. Baltimore: Johns Hopkins University Press.

Thick with Whales
Palumbi, S. R., and J. Roman. 2007. The history of whales read from DNA. In *Whales, Whaling, and Ocean Ecosystems*, Estes et al., eds., 102–115.

Roberts, C. 2007. *The Unnatural History of the Sea*. Washington, D.C.: Island Press.

Tuck, J. A. and R. Grenier. 1989. *Red Bay, Labrador: World Whaling Capital A.D. 1550–1600*. St. John's, Newfoundland: Atlantic Archaeology.

Passage of Plastic

Derraik, J. G. B. 2002. The pollution of the marine environment by plastic debris: A review. *Marine Pollution Bulletin* 44:842–852.

Krümmel, E. M., R. W. Macdonald, L. E. Kimpe, et al. 2003. Delivery of pollutants by spawning salmon. *Nature* 425:255–266.

Moore, C. J., S. L. Moore, M. K. Leecaster, et al. 2001. A comparison of plastic and plankton in the North Pacific Central Gyre. *Marine Pollution Bulletin* 42:1297–1300.

Rios, L. M., C. Moore, and P. R. Jones. 2007. Persistent organic pollutants carried by synthetic polymers in the ocean environment. *Marine Pollution Bulletin* 54:1230–1237.

United Nations Environment Programme, Global Programme of Action for the Protection of the Marine Environment from Land-based Activities (UNEP/GPA). 2006. *The State of the Marine Environment: Trends and Processes*. The Hague: UNEP/GPA.

Whaling Endangers More Than Whales

Alter, S. E., E. Rynes, and S. R. Palumbi. 2007. DNA evidence for historic population size and past ecosystem impacts of gray whales. *Proceedings of the National Academy of Science* 104:15162–15167.

Grebmeier, J. M., and N. M. Harrison. 1992. Seabird feeding on benthic amphipods facilitated by gray whale activity in the northern Bering Sea. *Marine Ecology Progress Series* 80:125–133.

Rouse, G. W., S. K. Goffredi, and R. C. Vrijenhoek. 2004. *Osedax*: Bone-eating marine worms with dwarf males. *Science* 305:668–671.

Smith, C. R. 2007. Bigger is better: The role of whales as detritus in marine ecosystems. In *Whales, Whaling, and Ocean Ecosystems*, Estes et al., eds., 286–302.

Smith, C. R., and A. R. Baco. 2003. Ecology of whale falls at the deep-sea floor. *Oceanography and Marine Biology: An Annual Review* 41:311–354.

Back from the Edge of Extinction

Heppell, S. S., P. M. Burchfield, and L. J. Peña. 2007. Kemp's ridley recovery: How far have we come, and where are we headed? In *Ridley Sea Turtles*, Plotkin, ed., 325–335.

Lewison, R. L., L. B. Crowder, and D. J. Shaver. 2003. The impact of turtle excluder devices and fisheries closures on loggerhead and Kemp's ridley strandings in the western Gulf of Mexico. *Conservation Biology* 17:1089–1097.

National Marine Fisheries Service, Kemp's ridley sea turtle.

Plotkin, P. T. 2007. Near extinction and recovery. In *Ridley Sea Turtles*, Plotkin, ed., 337–339.

Whither the Right Whale?

Kraus, S. D., M. W. Brown, H. Caswell, et al. 2005. North Atlantic right whales in crisis. *Science* 309:561–562.

Kraus, S. D. and R. M. Rolland. 2007. Right whales in the urban ocean. In *The Urban Whale: North Atlantic Right Whales at the Crossroads*, S. D. Kraus and R. M. Rolland, eds., 1–38. Cambridge: Harvard University Press.

———. 2007. The urban whale syndrome. In *Urban Whale*, Kraus and Rolland, eds., 488-513.

Myers, R. A., S. A. Boudreau, R. D. Kenney, et al. 2007. Saving endangered whales at no cost. *Current Biology* 17:R10–R11.

National Marine Fisheries Service. 2006. *Review of the Status of the Right Whales in the North Atlantic and North Pacific Oceans*. Silver Spring, Md.: National Marine Fisheries Service, Office of Protected Resources.

Vanderlaan, A. S. M., and C. T. Taggart. 2007. Vessel collisions with whales: The probability of lethal injury based on vessel speed. *Marine Mammal Science* 23:144–156.

7. EDGE OF CONTINENTS

Erlandson, J. M., M. H. Graham, B. J. Bourque, et al. 2007. The kelp highway hypothesis: Marine ecology, the coastal migration theory, and the peopling of the Americas. *Journal of Island and Coastal Archaeology* 2:161–174.

Food and Agricultural Organization of the United Nations. 2007. *State of World Fisheries and Aquaculture 2006*. Part 1, *World Review of Fisheries and Aquaculture*. Rome: Food and Agriculture Organization of the United Nations, FAO Fisheries and Aquaculture Department.

Marean, C. W., M. Bar-Matthews, J. Bernatchez, et al. 2007. Early human use of marine resources and pigment in South Africa during the Middle Pleistocene. *Nature* 449:905–909.

Window into the Past

Marean et al. Marine resources and pigment in South Africa.

McBrearty, S., and C. Stringer. 2007. Palaeoanthropology: The coast in color. *Nature* 449:793–794.

Pinnegar, J. K., and G. H. Engelhard. 2008. The 'shifting baseline' phenomenon: A global perspective. *Reviews in Fish Biology and Fisheries* 18:1–16.

Sea of Plenty

Leavenworth, W. B. n.d. The changing landscape of maritime resources in seventeenth-century New England. Gulf of Maine Cod Project, University of New Hampshire, unpublished ms.

———. 2007. Maine county fisheries, 1887–1889. Gulf of Maine Cod Project and History of Marine Animals Populations, University of New Hampshire. Personal communication, November 1.

Maine Mining and Industrial Journal. 1884. Swordfish off the coast of Maine. July 25: 5, 8.

Martin, S. J. 1881. Notes on the fisheries of Gloucester, Massachusetts. *Bulletin of the United States Fish Commission*, April 30.

Pinnegar and Engelhard, 'Shifting baseline' phenomenon.

Roberts, *Unnatural History*.

Snow, D. R. 1972. Rising sea level and prehistoric cultural ecology in northern New England. *American Antiquity* 37:211–221.

Empty Ocean

Ames, E. P. 2004. Atlantic cod structure in the Gulf of Maine. *Fisheries* 29:10–28.

Devine, J. A., K. D. Baker, and R. L. Haedrich. 2006. Deep-sea fishes qualify as endangered. *Nature* 439:29.

Gianni, M. 2004. *High Seas Bottom Trawl Fisheries and Their Impacts on the Biodiversity of Vulnerable Deep-Sea Ecosystems: Options for International Action*. Gland, Switzerland: World Conservation Union (IUCN).

Leavenworth, Changing landscape of maritime resources.

Rosenberg, A.A., W. J. Bolster, K. E. Alexander, et al. 2005. The history of ocean resources: Modeling cod biomass using historical records. *Frontiers in Ecology and the Environment* 3:84–90.

United Nations Environment Programme, *Environment for Development*.

Disappearing Giants of the Sea

Abercrombie, D. L., S. C. Clarke, and M. S. Shivji. 2005. Global-scale genetic identification of hammerhead sharks: Application to assessment of the international fin trade and law enforcement. *Conservation Genetics* 6:775–788.

Adams, W. F., S. L. Fowler, P. Charvet-Almeida, et al. 2006. *Pristis pectinata. In 2007 IUCN Red List of Threatened Species*. www.iucnredlist.org/search/details.php/18175/all.

Baum, J. K., and R. A. Myers. 2004. Shifting baselines and the decline of pelagic sharks in the Gulf of Mexico. *Ecology Letters* 7:135–145.

Charvet-Almeida, P., V. Faria, M. Furtado, et al. 2007. *Pristis perotteti*. In *2007 IUCN Red List of Threatened Species*. www.iucnredlist.org/search/details.php/18176/all.

Christensen, V., S. Guenette, J. J. Heymans, et al. 2003. Hundred-year decline of North Atlantic predatory fishes. *Fish and Fisheries* 4:1–24.

Jennings, S., and J. D. Blanchard. 2004. Fish abundance with no fishing: Predictions based on macroecological theory. *Journal of Animal Ecology* 73:632–642.

MacKenzie, B. R., and R. A. Myers. 2007. The development of the northern European fishery for north

Atlantic bluefin tuna *Thunnus thynnus* during 1900–1950. *Fisheries Research* 87:339–230.

Martins, C., and C. Knickle. n.d. Megamouth shark. *Icthyology at the Florida Museum of Natural History*. www.flmnh.ufl.edu/fish/Gallery/descript/Megamouth/Megamouth.htm.

Morato, T., R. Watson, T. J. Pitcher, et al. 2006. Fishing down the deep. *Fish and Fisheries* 7:24–34.

Musick, J. 2005. Introduction. In *Sharks, Rays, and Chimaeras: The Status of Chondrichthyan Fishes*, S. L. Fowler, R. D. Cavanaugh, M. Camhi, et al., eds., 1–3. IUCN/SSC Shark Specialist Group. Gland, Switzerland: World Conservation Union (IUCN).

Myers, R. A., J. K. Baum, T. D. Shepherd, et al. 2007. Cascading effects of the loss of apex predatory sharks from a coastal ocean. *Science* 315:1846–1850.

Myers, R. A., and B. Worm. 2003. Rapid worldwide depletion of predatory fish communities. *Nature* 423:280–283.

———. 2005. Extinction, survival or recovery of large predatory fishes. *Philosophical Transactions of the Royal Society B* 360:13–20.

Pauly, D., J. Alder, A. Bakun, et al. 2005. Marine fisheries systems. In *Ecosystems and Human Well-Being: Current State and Trends*, R. Hassan, R. Scholes, and N. Ash, eds., 477–511. Millennium Ecosystem Assessment Series, vol. 1. Washington, D.C.: Island Press.

Quattro, J. M., D. S. Stoner, W. B. Driggers, et al. 2006. Genetic evidence of cryptic speciation within hammerhead sharks (Genus *Sphyrna*). *Marine Biology* 148:1143–1155.

Extinction
LeBoeuf, B. J., K. W. Kenyon, and B. Villa-Ramirez. 1986. The Caribbean monk seal is extinct. *Marine Mammal Science* 2:70–72.

Myers et al., Cascading effects.

Pauly et al., Marine fisheries systems.

Prideaux et al., Mammalian responses.

Roberts, *Unnatural History*.

Sala, E., and N. Knowlton. 2006. Global marine biodiversity trends. *Annual Review of the Environment and Resources* 31:93–122.

United Nations Environment Programme, *Environment for Development*.

Wilson, *The Creation*.

Cascading Losses
Brashares, J. S., P. Arcese, M. K. Sam, et al. 2004. Bushmeat hunting, wildlife declines, and fish supply in West Africa. *Science* 306:1180–1183.

Frank, K. T., B. Petrie, J. S. Choi, et al. 2005. Trophic cascades in a formerly cod-dominated ecosystem. *Science* 308:1621–1623.

Jackson, J. B. C., M. X. Kirby, W. H. Berger, et al. 2001. Historical overfishing and the recent collapse of coastal ecosystems. *Science* 293:629–638.

Lynam, C. P., M. J. Gibbons, B. E. Axelsen, et al. 2006. Jellyfish overtake fish in a heavily fished ecosystem. *Current Biology* 16:R492–R493.

National Research Council, Ocean Studies Board. 2002. *Effects of Trawling and Dredging on Seafloor Habitat*. Washington, D.C.: National Academy Press.

United Nations Environment Programme, *Environment for Development*.

Farming the Sea
Agardy, T., J. Alder, P. Dayton, et al. 2005. Coastal systems. In *Ecosystems and Human Well-Being*, Hassan, Scholes, and Ash, eds., 515–549.

Ford, J. S., and R. A. Myers. 2008. A global assessment of salmon aquaculture impacts on wild salmonids. *PLoS Biology* 6:e33, doi:10.1371/journal.pbio.0060033.

Krkošek, M., J. S. Ford, A. Morton, et al. 2007. Declining wild salmon populations in relation to parasites from farm salmon. *Science* 318:1772–1775.

Naylor, R. L., R. J. Goldburg, J. H. Primavera, et al. 2000. Effect of aquaculture on world fish supplies. *Nature* 405:1017–1024.

Pauly et al., Marine fisheries systems.

Rosenberg, A. A. 2008. The price of lice. *Nature* 451:23–24.

A Future for Wild Fish
Barot, S., M. Heino, L. O'Brien, et al. 2004. Long-term trend in the maturation reaction norm of two cod stocks. *Ecological Applications* 14:1257–1271.

Berkeley, S. A., C. Chapman, and S. M. Sogard. 2004. Maternal age as a determinant of larval growth and survival in a marine fish, *Sebastes melanops*. *Ecology* 85:1258–1264.

Berkeley, S. A., M. A. Hixon, R. J. Larson, et al. 2004. Fisheries sustainability via protection of age structure and spatial distribution of fish populations. *Fisheries* 29:23–32.

Brodziak, J., and M. Traver. 2006. Haddock. In *Status of Fishery Resources off the Northeastern US*. Woods Hole, Mass.: Northeast Fisheries Science Center. www.nefsc.noaa.gov/sos/spsyn/pg/haddock/#top.

Food and Agricultural Organization, *State of World Fisheries and Aquaculture 2006*.

Hutchings, J. A. 2004. The cod that got away. *Nature* 428:899–935.

Hutchings, J. A., and J. D. Reynolds. 2004. Marine fish population collapses: Consequences for recovery and extinction risk. *Bioscience* 54:297–310.

Murawksi, S. A., R. Brown, H. L. Lai, et al. 2000. Large-scale closed areas as a fishery-management tool in temperate marine systems: The Georges Bank experience. *Bulletin of Marine Science* 66:775–798.

Palumbi, S. 2004. Why mothers matter. *Nature* 430:621–622.

United Nations Environment Programme, *Environment for Development*.

Worm, B., E. B. Barbier, N. Beaumont, et al. 2006. Impacts of biodiversity loss on ocean ecosystem services. *Science* 314:787–790.

8. RHYTHMS ON A REEF
Hata, H., and M. Kato. 2006. A novel obligate cultivation mutualism between damselfish and *Polysiphonia* algae. *Biology Letters* 2:593–596.

Patek, S. N., W. L. Korff, and R. I. Caldwell. 2004. Deadly strike mechanism of a mantis shrimp. *Nature* 428:819.

Sumich and Morrissey, 2004. *Biology of Marine Life*.

A Reef to Rival a Rainforest
Bruckner, A. W. 2002. Life-saving products from coral reefs. *Issues in Science and Technology* 18 (3): 39–44.

Levy, O., L. Appelbaum, W. Leggat, et al. 2007. Light-responsive cryptochromes from a simple multicellular animal, the coral *Acropora millepora*. *Science* 318:467–470.

Sala and Knowlton, Global marine biodiversity trends.

Stone, R. 2007. A world without corals? *Science* 316:678–681.

United Nations Environment Programme, *Environment for Development*.

Reefs through Time
Johnson, C. C. 2002. The rise and fall of rudist reefs. *American Scientist* 90:148–154.

Wood, R. 1999. *Reef Evolution*. Oxford: Oxford University Press.

Kelp in the Tropics
Graham, M. H., B. P. Kinlan, L. D. Druehl, et al. 2007. Deep-water kelp refugia as potential hotspots of tropical marine diversity and productivity. *Proceedings of the National Academy of Sciences* 104:16576–16580.

Miller, K. A., L. Garske, and G. Edgar. 2007. *Eisenia galapagensis*. In *2007 IUCN Red List of Threatened Species*. www.iucnredlist.org/search/details.php/63598/all.

Cool Coral
Roberts, J. M., A. J. Wheeler, and A. Freiwald. 2006. Reefs of the deep: The biology and geology of cold-water coral ecosystems. *Science* 312:543–546.

Fishing a Reef
Balmford, A., P. Gravestock, N. Hockley, et al. 2004. The worldwide costs of marine protected areas. *Proceedings of the National Academy of Sciences* 101:9694–9697.

Dulvy, N. K. 2006. Conservation biology: Strict marine protected areas prevent reef shark declines. *Current Biology* 16:R989–R990.

Fosså, J. H., P. B. Mortensen, and D. M. Furevik. 2002. The deep-water coral *Lophelia pertusa* in Norwegian waters: Distribution and fishery impacts. *Hydrobiologia* 471:1–12.

287

Freiwald, A., J. H. Fosså, A. Grehan, et al. 2004. *Cold-water Coral Reefs*. Cambridge: United Nations Environment Programme–World Conservation Monitoring Centre (UNEP-WCMC).

Friedlander, A. M., and E. E. DeMartini. 2002. Contrasts in density, size, and biomass of reef fishes between the northwestern and the main Hawaiian islands: The effects of fishing down apex predators. *Marine Ecology Progress Series* 230:23–264.

Reed, J. K., A. N. Shepard, C. C. Koenig, et al. 2005. Mapping, habitat characterization, and fish surveys of the deep-water *Oculina* coral reef Marine Protected Area: A review of historical and current research. In *Cold-water Corals and Ecosystems*, A. Freiwald and J. M. Roberts, eds., 443–465. Berlin: Heidelberg.

Roberts, Wheeler, and Freiwald, Reefs of the deep.

Sandin, S. A., J. E. Smith, E. E. DeMartini, et al. 2008. Baselines and degradation of coral reefs in the Northern Line Islands. *PLoS Biology* 6:e54, doi:10.1371/journal.pbio.0060054.

Stokstad, E. 2005. Alaskan coral preserved. *Science* 307:1027.

United Nations Environment Programme, *Environment for Development*.

The Waters Nearby

Gardner, T. A., I. M. Côté, J. A. Gill, et al. 2003. Long-term region-wide declines in Caribbean corals. *Science* 301:958–960.

Hughes, T. P., A. H. Baird, D. R. Bellwood, et al. 2003. Climate change, human impacts, and the resilience of coral reefs. *Science* 301:929–933.

Knowlton, N. 2001. The future of coral reefs. *Proceedings of the National Academy of Sciences* 98:5419–5425.

Palumbi, S. R. 2005. Germ theory for ailing corals. *Nature* 434:713–715.

Pandolfi, J. M., J. B. C. Jackson, N. Baron, et al. 2005. Are U.S. coral reefs on the slippery slope to slime? *Science* 307:1725–1726.

Parry, M. L., O. F. Canziani, J. P. Palutikof, et al., eds. 2007. Cross-chapter case studies. In *Climate Change 2007: Impacts, Adaptation and Vulnerability. Contribution of Working Group II to the Fourth Assessment Report of the Intergovernmental Panel on Climate Change*, 843–868. Cambridge: Cambridge University Press.

Stone, World without corals?

Warming Sea

Aronson, R. B., and W. F. Precht. 2006. Conservation, precaution, and Caribbean reefs. *Coral Reefs* 25: 441–450.

Aronson, R. B., W. F. Precht, I. G. Macintrye, et al. 2000. Ecosystems: Coral bleach-out in Belize. *Nature* 405:36.

Fischlin, A. G. F. Midgley, J. T. Price, et al. 2007. Ecosystems, their properties, goods, and services. In *Climate Change 2007,* Parry et al., eds., 211–272.

Gardner et al., Region-wide declines in Caribbean corals.

Graham, N. A. J., S. K. Wilson, S. Jennings, et al. 2006. Dynamic fragility of oceanic coral reef ecosystems. *Proceedings of the National Academy of Sciences* 103:8425–8429.

Hoegh-Guldberg, O., P. J. Mumby, A. J. Hooten, et al. 2007. Coral reefs under rapid climate change and ocean acidification. *Science* 318:1737–1742.

IPCC. 2007. Summary for policymakers. In *Climate Change 2007,* Parry et al., eds., 7–22.

Parry et al., Cross-chapter case studies.

Stone, World without corals?

Safe Harbor

Erwin, D. 2001. Lessons from the past: Biotic recoveries from mass extinctions. *Proceedings of the National Academy of Sciences* 98:5399–5403.

Freiwald et al. *Cold-water Coral Reefs*.

Hoegh-Guldberg et al., Coral reefs under rapid climate change.

Hughes et al., Climate change, human impacts, and the resilience of coral reefs.

Kano, A., T. G. Ferdelman, T. Williams, et al. 2007. Age constraints on the origin and growth history of a deep-water coral mound in the northeast Atlantic drilled during Integrated Ocean Drilling Program Expedition 307. *Geology* 35:1051–1054.

Schuttenberg, H., and O. Hoegh-Guldberg. 2007. A world with corals: What would it take? *Science* 318:42–44.

9. FAR ENDS OF EARTH

Anisimov, O. A., D. G. Vaughan, T. V. Callaghan, et al. 2007. Polar regions (Arctic and Antarctic). In *Climate Change 2007,* Parry et al., eds., 653–685.

Laidre, K. L., M. P. Heide-Jørgensen, R. Dietz, et al. 2003. Deep-diving by narwhals *Monodon monoceros*: Differences in foraging behavior between wintering areas? *Marine Ecology Progress Series* 261:269–281.

Loeng, H., K. Brander, E. Carmack, et al. 2005. Marine systems. In *Arctic Climate Impact Assessment Scientific Report*, C. Symon, L. Arris, and B. Heal, eds., 453–538. Cambridge: Cambridge University Press.

Nweeia, M. T., N. Eidelman, F. C. Eichmiller, et al. 2005. Hydrodynamic sensor capabilities and structural resilience of the male narwhal tusk. Paper presented at the 16th Biennial Conference on the Biology of Marine Mammals, San Diego, December 13.

Cold and Ice

Anisimov et al. Polar regions.

Cziko, P. A., C. W. Evans, C.-H. C. Cheng, et al. 2006. Freezing resistance of antifreeze-deficient larval Antarctic fish. *Journal of Experimental Biology* 209: 407–420.

Fischlin et al. Ecosystems, their properties, goods, and services.

Gilbert, C., Y. L. Maho, M. Perret, et al. 2007. Body temperature changes induced by huddling in

breeding male emperor penguins. *American Journal of Physiology—Regulatory, Integrative and Comparative Physiology* 292:R176–R185.

Kelly, B. P. 2001. Climate change and ice breeding pinnipeds. In *"Fingerprints" of Climate Change: Adapted Behaviour and Shifting Species Ranges*, G.-R. Walter, C. A. Burga, and P. J. Edwards, eds., 43–55. New York: Kluwer Academic/Plenum.

Loeng et al., Marine systems.

Sumich and Morrissey, *Biology of Marine Life*.

Of Ice and Men in the Northwest Passage

Anisimov et al., Polar regions.

Delaney, J. 2004. Of maps and men: In pursuit of a northwest passage. libweb5.princeton.edu/visual_materials/maps/websites/northwest-passage/titlepage.htm.

Kerr, R. 2007. Is battered Arctic sea ice down for the count? *Science* 318:33–34.

Reid, P. C., D. G. Johns, M. Edwards, et al. 2007. A biological consequence of reducing Arctic ice cover: Arrival of the Pacific diatom *Neodenticula seminae* in the North Atlantic for the first time in 800,000 years. *Global Change Biology* 13:1910–1921.

Stroeve, J., M. Serreze, S. Drobot, et al. 2008. Arctic sea ice extent plummets in 2007. *Eos* 89:13–14.

Plumbing the Depths

Brandt, A., A. J. Gooday, S. N. Brandão, et al. 2007. First insights into the biodiversity and biogeography of the Southern Ocean deep sea. *Nature* 447:307–311.

Brinkhuis, H., S. Schouten, M. E. Collinson, et al. 2006. Episodic fresh surface waters in the Eocene Arctic Ocean. *Nature* 441:606–609.

Jakobsson, M., J. Backman, B. Rudels, et al. 2007. The early Miocene onset of a ventilated circulation regime in the Arctic Ocean. *Nature* 447:986–989.

Sluijs, A., S. Schouten, M. Pagani, et al., 2006. Subtropical Arctic Ocean temperatures during the Palaeocene/Eocene thermal maximum. *Nature* 441:610–613.

Melting

Comiso, J. C., C. L. Parkinson, R. Gersten, et al. 2008. Accelerated decline in the Arctic sea ice cover. *Geophysical Research Letters* 35, L01703, doi:10.1029/2007G1031972.

Kerr, Is battered Arctic sea ice down?

Maslanik, J. A., C. Fowler, J. Stroeve, et al. 2007. A younger, thinner Arctic ice cover: Increased potential for rapid, extensive, sea-ice loss. *Geophysical Research Letters* 34, L24501, doi:10.1029/2007GL032043.

Ngheim, S. V., I. G. Rigor, D. K. Perovich, et al. 2007. Rapid reduction of Arctic perennial sea ice. *Geophysical Research Letters* 34, L19504, doi:10.1029/2007GL 031138.

Overland, J. E., and M. Wang. 2007. Future regional Arctic sea ice declines. *Geophysical Research Letters* 34, L17705, doi:10.1029/2007GL030808.

Serreze, M.C., M. M. Holland, and J. Stroeve. 2007. Perspectives on the Arctic's shrinking sea-ice cover. 2007. *Science* 315:1533–1536.

Stroeve, J., M. M. Holland, W. Meier, et al. 2007. Arctic sea ice decline: Faster than forecast. *Geophysical Research Letters* 34, L09501, doi:10.1029/2007GL029703.

Stroeve et al., Arctic sea ice extent plummets.

Feeling the Heat

Amstrup, S. C., B. G. Marcot, and D. C. Douglas. 2007. Forecasting the range-wide status of polar bears at selected times in the 21st century. Reston, Va.: U.S. Department of the Interior, U.S. Geological Survey.

Cooper, L. W., C. J. Ashjian, S. L. Smith, et al. 2006. Rapid seasonal sea-ice retreat in the Arctic could be affecting Pacific Walrus (*Odobenus rosmarus divergens*) recruitment. *Aquatic Mammals* 32:98–102.

Ducklow, H. W., K. Baker, B. G. Martinson, et al. 2007. Marine pelagic ecosystems: The West Antarctic Peninsula. *Philosophical Transactions of the Royal Society B* 362:67–94.

Fischbach, A. S., S. C. Amstrup, and D. C. Douglas. 2007. Landward and eastward shift of Alaskan polar bear denning associated with recent sea ice changes. 2007. *Polar Biology* 30:1395–1405.

Grebmeier, J. M., J. E. Overland, S. E. Moore, et al. 2006. A major ecosystem shift in the northern Bering Sea. *Science* 311:1461–1464.

Kooyman, G. L., D. G. Ainley, G. Ballard, et al. 2007. Effects of giant icebergs on two emperor penguin colonies in the Ross Sea, Antarctica. *Antarctic Science* 19:31–38.

Loeng et al., Marine systems.

Monnet, C., and J. S. Gleason. 2006. Observations of mortality associated with extended open-water swimming by polar bears in the Alaskan Beaufort Sea. *Polar Biology* 29:681–687.

Stirling, I., and C. L. Parkinson. 2006. Possible effects of climate warming on selected populations of polar bears (*Ursus maritimus*) in the Canadian Arctic. *Arctic* 59:261–275.

Stokstad, E. 2007. Boom and bust in a polar hot zone. *Science* 315:1522–1523.

A Culture at Risk

ACIA. 2004. *Impacts of a Warming Arctic: Arctic Climate Impact Assessment*. Cambridge: Cambridge University Press.

George, J. C., H. P. Huntington, K. Brewster, et al. 2004. Observations on shorefast ice dynamics in Arctic Alaska and the responses of the Iñupiat hunting community. *Arctic* 57:363–374.

Krupnik, I., and D. Jolly, eds. 2002. *The Earth Is Faster Now: Indigenous Observations of Arctic Environmental Change*. Fairbanks, Alaska: Arctic Research Consortium of the United States.

Laidler, G. J., and T. Ikummaq. 2008. Human geographies of sea ice: Freeze/thaw processes around Igloolik, Nunavut, Canada. *Polar Record* 44:127–153.

National Museum of Natural History, Smithsonian Institution. n.d. Arctic: A friend acting strangely. www.forces.si.edu/arctic.

Beyond the Ice

Ainley, D., G. Ballard, S. Ackley, et al. 2007. Paradigm lost, or is top-down forcing no longer significant in the Antarctic marine ecosystem? *Antarctic Science* 19:283–290.

Atkinson, A., V. Siegel, E. Pakhomov, et al. 2004. Long-term decline in krill stock and increase in salps within the Southern Ocean. *Nature* 432:100–103.

Boyd, P. W. 2007. Iron findings. *Nature* 446:989–990.

Intergovernmental Panel for Climate Change (IPCC). 2007. Summary for policymakers. In *Climate Change 2007: Mitigation. Contribution of Working Group III to the Fourth Assessment Report of the Intergovernmental Panel on Climate Change*, B. Metz, O. R. Davidson, P. R. Bosch, et al., eds., 1–24. Cambridge: Cambridge University Press.

Kock, K.-H., K. Reid, J. Croxall, et al. 2007. Fisheries in the Southern Ocean: An ecosystem approach. *Philosophical Transactions of the Royal Society B* 362:2333–2349.

Lutz, M. J., K. Caldeira, R. B. Dunbar, et al. 2007. Seasonal rhythms of net primary production and particulate organic carbon flux to depth describe the efficiency of biological pump in the global ocean. *Journal of Geophysical Research* 112, C10011, doi:10.1029/2006JC003706.

Morato et al., Fishing down the deep.

Myers and Worm, Rapid worldwide depletion.

Reverberations

Bindschadler, R. 2006. Hitting the ice sheets where it hurts. *Science* 311:1720–1721.

Hanna, E., P. Huybrechts, K. Steffen, et al. 2008. Increased runoff from melt from the Greenland ice sheet: A response to global warming. *Journal of Climate* 21:331–341.

Intergovernmental Panel for Climate Change (IPPC). 2007. Summary for policymakers. In *Climate Change 2007: Synthesis Report*. www.ipcc.ch/pdf/assessment-report/ar4/syr/ar4_syr_spm.pdf.

Overpeck, J. T., B. L. Otto-Bliesner, G. H. Miller, et al. 2006. Paleoclimatic evidence for future ice-sheet instability and rapid sea-level rise. *Science* 311:1747–1750.

Rahmstorf, S. 2007. A semi-empirical approach to projecting future sea-level rise. *Science* 315:368–370.

Rahmstorf, S., A. Cazenave, J. A. Church, et al. 2007. Recent climate observations compared to projections. *Science* 316:709.

Rignot, E. 2008. Research highlights: Southern melt. *Nature* 451:226.

Rignot, E., and P. Kanagaratnam. 2006. Changes in the velocity structure of the Greenland ice sheet. *Science* 311:986–990.

Rohling, E. J., K. Grant, Ch. Hemleben, et al. 2008. High rates of sea-level rise during the last interglacial period. *Nature Geoscience* 1:38–42.

Shepherd, A., and D. Wingham. 2007. Recent sea-level contributions of the Antarctic and Greenland ice sheets. *Science* 315:1529–1531.

Tedesco, M. 2007. A new record in 2007 for melting in Greenland. *Eos* 88:383.

Velicogna, I., and J. Wahr. 2006. Measurements of time-variable gravity show mass loss in Antarctica. *Science* 311:1754–1756.

10. WHERE RIVERS MEET THE SEA

Newcomers

Hobbs, C. H., III. 2004. Geological history of Chesapeake Bay, USA. *Quaternary Science Reviews* 23:641–661.

Long, *Rise of Fishes*.

Wetmore, A. 1926. Observations on fossil birds described from the Miocene of Maryland. *Auk* 43: 462–468.

Rimmed by Salt Marsh

Castro, J. I. 1993. The shark nursery of Bulls Bay, South Carolina, with a review of the shark nurseries of the southeastern coast of the United States. *Environmental Biology of Fishes* 38:37–48.

Rimmed by Mangrove

Mumby, P. J., A. J. Edwards, J. E. Arias-González, et al. 2004. Mangroves enhance the biomass of coral reef fish communities in the Caribbean. *Nature* 427:533–536.

Wong, P. P. 2005. The coastal environment of Southeast Asia. *In The Physical Geography of Southeast Asia*, A. Gupta, ed., 177–192. Oxford: Oxford University Press.

Chesapeake and Beyond

Agardy et al., 2005. Coastal systems.

Bricker, S., B. Longstaff, W. Dennison, et al. 2007. *Effects of Nutrient Enrichment in the Nation's Estuaries: A Decade of Change*. NOAA Coastal Ocean Program Decision Analysis Series No. 26. Silver Spring, Md.: National Centers for Coastal Ocean Science.

Chesapeake Bay Foundation. 2007. Bad waters: Dead zones, algal blooms, and fish kills in the Chesapeake Bay region in 2007. www.cbf.org/site/DocServer/CBF_BadWatersReport.pdf?docID=10003.

Hariot, T. 2007 [1590]. *A Briefe and True Report of the New Found Land of Virginia*. Charlottesville: University of Virginia Press.

Hulton, P. 1984. *America, 1585: The Complete Drawings of John White*. Chapel Hill: University of North Carolina Press.

Kane, A. S., C. B. Stine, L. Hungerford, et al. 2007. Mycobacteria as environmental portent in Chesapeake Bay fish species. *Emerging Infectious Diseases* 13:329–331.

Kemp, W. M., W. R. Boynton, J. E. Adolf, et al. 2005. Eutrophication of Chesapeake Bay: Historical trends and ecological interactions. *Marine Ecology Progress Series* 303:1–29.

Lotze, H. K., H. S. Lenihan, B. J. Bourque, et al. 2006. Depletion, degradation, and recovery potential of estuaries and coastal seas. *Science* 312:1806–1809.

Mead, J. G., and E. D. Mitchell. 1984. Atlantic gray whales. In *The Gray Whale: Eschrichtius robustus*, M. L. Jones, S. L. Swartz, and S. Leatherwood, eds., 33–53. Orlando: Academic Press.

Selman, M., S. Greenhalgh, R. Diaz, et al. 2008. Eutrophication and hypoxia in coastal areas: A global assessment of the state of knowledge. *WRI Policy Note, Water Quality: Eutrophication and Hypoxia*, no. 1, March.

Stoecker, D. K., J. E. Adolf, A. R. Place, et al. 2008. Effects of the dinoflagellates *Karlodinium veneficum* and *Prorocentrum minimum* on early life history stages of the eastern oyster (*Crassostrea virginica*). *Marine Biology* 154:81–90.

Thomas, P., M. S. Rahman, I. A. Kahn, et al. 2007. Widespread endocrine disruption and reproductive impairment in an estuarine fish population exposed to seasonal hypoxia. *Proceedings of the Royal Society B* 274:2693–2701.

United Nations Environment Programme, *Environment for Development*.

Domestication of Estuaries and Coastal Waters

Agardy et al., Coastal systems.

Cooley, H., P. H. Gleick, and G. Wolff. 2006. Desalination, with a grain of salt: A California perspective. Oakland: Pacific Institute for Studies in Development, Environment, and Security.

Mumby et al., Mangroves enhance the biomass of coral reef fish communities.

Naylor, R. L., R. J. Goldburg, H. Mooney, et al. Ecology: Nature's subsidies to shrimp and salmon farming. *Science* 282:883–884.

Pauly et al., Marine fisheries systems.

Ruiz, G. M., and D. F. Reid, eds. 2007. Current state of understanding about the effectiveness of ballast water exchange (BWE) in reducing aquatic nonindigenous species (ANS) introductions to the Great Lakes Basin and Chesapeake Bay, USA: Synthesis and analysis of existing information. NOAA Technical Memorandum GLERL-142. Ann Arbor: NOAA Great Lakes Environmental Research Laboratory.

Semmens, B. X., E. R. Buhle, A. K. Salomon, et al. 2004. A hotspot of non-native marine fishes: Evidence for the aquarium trade as an invasion pathway. *Marine Ecology Progress Series* 266:239–244.

Whitfield, P. E., J. A. Hare, A. W. David, et al. 2007. Abundance estimates of the Indo-Pacific lionfish *Pterois volitans/miles* complex in the Western North Atlantic. *Biological Invasions* 9:53–64.

Rising Water

Colette, A. 2007. *Case Studies on Climate Change and World Heritage*. Paris: UNESCO World Heritage Centre.

Cronin, W. B. 2005. *The Disappearing Islands of the Chesapeake*. Baltimore: Johns Hopkins University Press.

Ericson, J. P., C. J. Vörösmarty, S. L. Dingman, et al. 2006. Effective sea-level rise and deltas: Causes of change and human dimension implications. *Global and Planetary Change* 50:63–82.

McGranahan, G., D. Balk, and B. Anderson. 2007. The rising tide: Assessing the risks of climate change and human settlements in low elevation coastal zones. *Environment and Urbanization* 19:17–37.

Mimura, N., L. Nurse, R. F. McLean, et al. 2007. Small islands. In *Climate Change 2007,* Parry et al., eds., 687–716.

Parry et al., Cross-chapter case studies.

Rahmstorf et al., Recent climate observations.

Stanley, J. D., F. Goddio, T. F. Jorstad, et al. 2004. Submergence of ancient Greek cities off Egypt's Nile delta—A cautionary tale. *GSA Today* 14:4–10.

OUR WATER, OUR WORLD

Halpern et al., Global map of human impact on marine ecosystems.

Knoll, *Life on a Young Planet*.

Mayhew, Jenkins, and Benton, Global temperature and biodiversity.

McGranahan, Balk, and Anderson, Rising tide.

Myers, N., and A. Knoll. 2001. The biotic crisis and the future of evolution. *Proceedings of the National Academy of Sciences* 98:5389–5392.

Pauly et al., Marine fisheries systems.

Royer, Berner, and Park, Climate sensitivity.

Srinivason, U. T., S. P. Carey, E. Hallstein, et al. 2008. The debt of nations and the distribution of ecological impacts from human activities. *Proceedings of the National Academy of Sciences* 105:1768–1773.

Wilson, *Future of Life*.

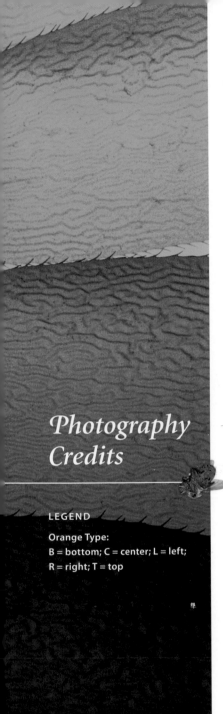

Photography Credits

LEGEND

Orange Type:
B = bottom; C = center; L = left;
R = right; T = top

INDEPENDENT PHOTOGRAPHERS

Jason Bradley, front endsheet; Eric Cheng, 165; Brandon Cole, 22, 81B, 118RB, 138LT, 138LB, 145LT, 145B, 171T, 174, 175; Feargus Cooney, 65BR; Chuck Davis, 193T; David Doubilet, 63, 115RT, 159, 194–195; Per Flood, Bathybiologica. no, 108L-1; Jürgen Freund, 237B; Stephen Frink, 191RT; Daisy Gilardini, 216B, 219; David Hall, 98LB, 139B, 189, 212L, 212R, 252; Howard Hall, 8–9, 14–15, 92T; Michele Hall, 12–13; Mauricio Handler, 213; Wolcott Henry, 119T; Richard Herrmann, 188T; Ralph Lee Hopkins, 16, 214–215; George H. H. Huey, 183; Chris Huss, 221; Raphael Lacoste, 88; Frans Lanting, 89L, 201LB, 216T; Dylan Lossie, 72–73; Richard A. Lutz, 26–27, 38; Ian R. MacDonald, Texas A & M University–Corpus Christi, 39B; John and Ann Mahan, 64; Alex Mustard, 50–51; Amos Nachoum, 222–223; Richard Olensius, 218; Michael Patrick O'Neill, 169R, 191RB, 240LT; Heather Perry, 6–7, 108B, 184–185; Lee Peterson, 276; Ian Plant, 239R; Enric Sala, 208; Kevin Schafer, 54–55T, 83R, 129RT; Johann Schumacher, 4–5; Andre Seale, 52–53; Douglas David Seifert, 60, 119BC, 191LB; Paul Sutherland, 186RT; Tom Till, 91; John Tobin, 238L; Masa Ushioda, 11; Staffan Widstrand, 154; Dwane Wilkin, 114; Art Wolfe, 246–247; Shuhai Xiao, 90RB

AGENCIES AND COLLECTIONS

Airphoto: Jim Wark, 68B

Alamy: Bill Bachman, 47; Andrew Bargery, 233B; blickwinkel: Hecker, 195B; Simon Bracken, 127T; Bryan and Cherry Alexander Photography, 179C; Rosen Dimitrov, 70–71; Frank Lane Picture Agency, 54–55RB; Galen Rowell/Mountain Light, 234–235; Gondwana Photo Art, 46; imagebroker: Bahnmueller, 69B; Jon Arnold Images, 134; LOOK Die Bildagentur der Fotografen GmbH: Per-Andre Hoffmann, 62TC; Nordicphotos: Sigurgeir Sigurjonsson, 43B; Randolph Images, 95; Westend 61: Martin Rietze, 61BC

Alaska Stock: Amanda Byrd, 223

Arco Images: K. Mosebach, 89B

ARKive: Sean Connell, 206T

Aurora: Christopher Herwig, 193C, 193B

Auscape: Leo McElfresh, 67T

Auscape/Minden Pictures: Jean-Paul Ferrero, 224; Reg Morrison, 126LT; D. Parer and E. Parer-Cook, 129B

Corbis: 42, 160–161; Theo Allofs, 65LB; Remi Benali, 135; Jonathan Blair, 147B; Ralph A. Clevenger, 280, 290, 292; DK Limited, 122; European Pressphoto Agency: Altaf Qadri, 150–151; European Pressphoto Agency: Dennis Sabangan, 61LB; Stephen Frink, 85; Godong: Thierry Brésillon, 133; Rob Howard, 131B; George H. H. Huey, 127B; Peter Johnson, 164; Bob Krist, back cover; Liu Liqun, 90T; Robert Matheson, 86–87; Joe McDonald, 130; David Muench, 142–143; Roger Ressmeyer, 131T, 137; Reuters: Rafiqur Rahman, 243T; Patrick Robert, 48; Kevin Schafer, 61RB; Paul A. Souders, 220T, 228–229; George Steinmetz, 200–201; Keren Su, 83L; Sygma: Pierre Vauthey, 54LB; WEDA/European Pressphoto Agency: 62TC; Steve Wilkings, 75; Zefa: Tobias Bernhard, 113R; Zefa: Jason Hosking, 123

Deep Sea Photography: Peter Batson, 29, 36LT, 36LB, 36RT, 37T, 37B, 39T, 40, 41, 98RL, 186LT

Digital Railroad: Kenneth Garrett, 180–181; Cindy Miller Hopkins, 132; Randy Santos, 232–233; Art Wolfe, 44

Futura-Sciences: Alexis Rosenfeld, 240RT

Getty Images: Daisy Gilardini, 76–77; Image Makers 62LT; Tim Laman, 236–237; G. Brad Lewis, 24–25; Fred Mayer, 82R; Sakis Papadopoulos, 243; Donovan Reese, 65; Eric Rorer, 128T; Melford Taylor, 238R; Atsushi Takeda, 124–125; Michael Townsend, 244–245; Stuart Westmorland, 198–199

Ifremer: Michel Segonzac, 36RB

Image Quest Marine: Peter Herring, 58RC; Peter Parks, 107B, 108L-2; Andre Seale, 119RB

Images & Stories: Zafer Kizilkaya, 207R

Images of Nature Stock: Thomas D. Mangelsen, back endsheet

Minden Pictures: Ingo Arndt, 106, 148B, 220C, 220B, 251; Fred Bavendam, 203T, 204; Jim Brandenburg, 67B; Matthias Breiter, 239L; Tui De Roy, 43T, 80L, 110, 128LT, 128LB, 129LT-1, 129LT-2; Michael Durham, 126LB; Frank Lane Picture Agency: D. P. Wilson, 108L-3; Niels Kooyman, 2–3; Frans Lanting, 99; Yva Momatiuk and John Eastcott, 171B; Chris Newbert, 278; Flip Nicklin, front cover, 81T, 104,167B, 172; Norbert Wu, 227

Monterey Bay Aquarium Research Institute: Karen Osborn, 107LT, 107LB

Muench Photography: David Muench, 82L

National Geographic Image Collection: Bill Curtsinger, 169L, 179B; David Doubilet, 203B, 230–231; Raymond Gehman, 45, 94; George Grall, 235; Tim Laman, 182, 241, 242; Paul Nicklen, 111, 170, 197R, 225, 226–227; Brian J. Skerry, 100–101, 191LT, 197L; Steve Winter, 168–169

National Museum of Natural History, Smithsonian Institution: Chip Clark, 66T, 66B, 68L, 69T, 92LB, 92CB, 92RB, 93 (all), 96L, 96R, 97LT, 97RT, 116R, 117, 139T-1, 139T-3, 139T-4, 139T-5, 145LB

Nature Picture Library: Georgette Douwma, 211B; Tony Evans, 149; Jürgen Freund, 115TL, 205R; Martha Holmes, 210–211; David Noton, 120; Constantinos Petrinos, 58RC; David Shale, 57LT, 57LC, 57RT, 57RC, 57RB, 58LT, 109

Nature Picture Library/Minden Pictures: David Noton, 113LB

New England Aquarium: Yan Guilbault, 179T; Owen Nichols, 177; Jessica Taylor, 178

Oceanwide Images: Gary Bell, 18–19, 84–85, 115B, 118LB, 167RT, 196, 201; Rudie Kuiter, 58RT

Origins Museum Institute: Marty Martin 116L

Oxford Scientific: Dr. Dennis Kunkel, 102–103RB

Photo Researchers, Inc.: James L. Amos, 141B; L. K. Broman, 138B; Steve Gschmeissner, 102LB; Sinclair Stammers, 140

Photoshot/Oceans Image: Charles Hood, 202

SeaPics.com: 58RB, 118BC, 119LB, 148LT, 148LB, 166–167, 173, 176, 188B, 190LB, 190RB, 192, 205L, 209LT, 209RT, 296; David Fleetham, 195T; Doug Perrine, 80R, 112–113; Masa Ushioda, 162–163

South Australian Museum: Jim Gehling, 89LC, 89RC

Splashdown Direct: 186B

Stephen Low Company: Emory Kristof, 28, 31, 32–33, 35; William Reeve and Emory Kristof, 33B

UWPhoto: Aage Jakobsen, 185B; Erling Svensen, 206L, 206R, 207L; Kare Telnes, 57LB

V & W/Image Quest Marine: Mark Conlin, 78–79

Vireo: Ted Daeschler, 121

Visuals Unlimited: Albert J. Copley, 141T; Wim van Egmond, 105; Dr. Dennis Kunkel, 102–103RB; Ken Lucas, 89R, 146, 147T; MedicalRF.com, 49; Mark A. Schneider, 139T-2

Woods Hole Oceanographic Institute: Larry Madin, 58LB, 108L-4

Index

Page numbers in **blue boldface** indicate photographs

Blue button hydrozoan, Porpita pacifica, *Hawaii.*

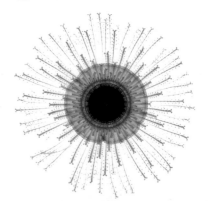